Praise for "genre-busting writer" (*Salon*)
Patricia Pearson and *Opening Heaven's Door*

"Readers will be humbled and filled with a sense of hope rather than fear as they realize that the deaths of loved ones, or even their own deaths, are not losses, but simply transitions. A fascinating and candid analysis of the process of dying."

—*Kirkus Reviews*

"This remarkable new book by Patricia Pearson is a rare thing: bringing journalistic rigor to an impossible question. . . . The book succeeds so well because it favors questions over answers, humility over certainty, and (when called for) crunchy ice-breaking humor over earnestness. But mostly it succeeds because of its unabashed concern with love as it's experienced not only by those at heaven's door but by the human tribe that's inevitably left behind when someone dies. Love, too, is a mystery that changes us."

—*The Globe and Mail*

"A wide-ranging account of discoveries and evolving understanding about life, death, the afterlife, and the true dimensions of consciousness. Numerous firsthand accounts, observations, and results of scientific research provide a readable primer on psi phenomena, significantly expanding our understanding of the realities of our existence."

—*Light of Consciousness*

"The word is out: you don't die when you die. That's the message from around 15 million Americans who have experienced a near-death experience, as Patricia Pearson, in sparkling prose, shows in this enormously engaging book. I know, I know: this premise causes serious intellectual indigestion in die-hard skeptics, but we should not be diverted by their leaky arguments. The fear of total annihilation with physical death has caused more suffering in human history than all the physical diseases combined. Pearson's message is a Great Cure for this Great Fear. This book conveys deep meaning and hope. It takes the pressure off and makes life more fulfilling and joyous. There is only one reason why you should *not* read this magnificent book: if you have a secret way not to die. But since the statistics so far are against you, let Pearson be your guide."

—Larry Dossey, MD, author of
*One Mind: How Our Individual
Mind Is Part of a Greater
Consciousness and Why It Matters*

"Pearson brings her blend of humor, sympathy, and keen critical intelligence to a topic that is all too often off-limits to writers of her caliber. This is exactly the smart book on the possibility of an afterlife that readers curious about the topic but leery of mush have been looking for."

—Ptolemy Tompkins, author of
The Modern Book of the Dead and
collaborator with Eben Alexander,
MD, on *Proof of Heaven*

"Pearson has brought us something rare: a unique blend of gifted storytelling combined with exhaustive scientific research about dying, grief, and spiritual connectivity. *Opening Heaven's Door* leaves us enthralled that death's mystery may be life's solution."

—Allan J. Hamilton, MD, author
of *The Scalpel and the Soul*

"Hardheaded and openhearted, Pearson has brought together riveting accounts of near-death experiences that will shake your assumptions about where life ends, and what death means. For seekers and skeptics alike, it is profoundly comforting, questing, and wise."

—Marni Jackson, author of
Pain: The Fifth Vital Sign

"In this compelling and provoking read, Patricia Pearson examines death and dying with uncommon thoughtfulness, asking questions too rarely asked. Moving and insightful, *Opening Heaven's Door* is an important work for all of us struggling with the inevitably of death."

—Steven Galloway, author
of *The Confabulist*

"Your life is over the moment you die. So I used to believe, with something like religious fervor. And then I read *Opening Heaven's Door*, and such is the power and art, the passion and rigor of Patricia Pearson's writing that I'm not nearly so sure of myself. This is a splendid book in all the ways a book can be splendid. It is a book to be read and re-read and urged upon friends."

—Barbara Gowdy, author of
We So Seldom Look on Love

"*Opening Heaven's Door* is a fascinating examination of the conclusion of all our struggles and victories: the instant of death. This omnivorous journey through grief and neuroscience and spirituality carries the reader swiftly along and into places we can never truly know—but Pearson provides an unprecedented glimpse."

—Kevin Patterson, author
of *Country of Cold*

Also by Patricia Pearson

FICTION

Playing House

Believe Me

NONFICTION

When She Was Bad: How and Why Women Get Away with Murder

Area Woman Blows Gasket

A Brief History of Anxiety—Yours and Mine

OPENING HEAVEN'S DOOR

What the Dying Are

Trying to Say About

Where They're Going

PATRICIA PEARSON

ATRIA PAPERBACK

NEW YORK • LONDON • TORONTO
SYDNEY • NEW DELHI

Note to Readers: Names and/or identifying details of some
of the people portrayed in this book have been changed.

ATRIA PAPERBACK
An Imprint of Simon & Schuster, Inc.
1230 Avenue of the Americas
New York, NY 10020

First Atria Paperback edition May 2015

ATRIA PAPERBACK and colophon are trademarks of Simon & Schuster, Inc.

For information about special discounts for bulk purchases,
please contact Simon & Schuster Special Sales at 1-866-506-1949
or business@simonandschuster.com.

The Simon & Schuster Speakers Bureau can bring authors to your
live event. For more information or to book an event contact the
Simon & Schuster Speakers Bureau at 1-866-248-3049
or visit our website at www.simonspeakers.com.

Designed by Paul Dippolito

Manufactured in the United States of America

10 9 8 7 6 5 4 3 2 1

Library of Congress Cataloging-in-Publication Data
Pearson, Patricia, date.
    Opening heaven's door : what the dying are saying about where they're going /
Patricia Pearson.—First [edition].
        pages cm
    Includes bibliographical references and index.
    1. Near-death experiences.   2. Future life.   I. Title.
    BF1045.N4P43 2014   133.901′3—dc23
2013045211

ISBN 978-1-4767-5706-3
ISBN 978-1-4767-5707-0 (pbk)
ISBN 978-1-4767-5708-7 (ebook)

For my family, and the tribe

"Blessed are those who mourn."

Contents

OPENING
HEAVEN'S
DOOR

CHAPTER 1

An Unexpected Vision

My father died in his blue-striped pajamas on a soft bed in a silent house. He wasn't ailing. At three or four in the morning, he gave out a sigh, loud enough to wake my mother, who sleepily assumed that he was having a bad dream. A sigh, a moan, a final breath escaping. She leaned over to rub his back, and then retreated into her own cozy haze of unconsciousness. Morning arrived a few hours later as a thin suffusion of northern March light. She roused herself and walked around the prone form of her husband of fifty-four years to go to the bathroom.

Downstairs to the humdrum rituals of the kitchen. Brewing coffee, easing her teased-apart English muffin halves into the toaster, listening to the radio, on which I was being interviewed about a brand-new book. Her youngest of five children, I was providing commentary about a lawsuit brought by a man who had suffered incalculable psychological damage from finding a dead fly in his bottle of water.

"Did he have grounds?" the host was asking me. Was it possible for a life to unravel at the prospect of one dead fly?

My mother spread her muffin with marmalade, thought ahead to her day. Some meetings, a luncheon, an outing with her granddaughter Rachel, who was visiting for March break. She didn't wonder why Geoffrey, my father, still remained in their bed. No heightened sense of vigilance for a healthy man who'd just turned eighty.

In families, attention is directed toward crisis, and during the early spring of 2008, we were all transfixed by my sister Katharine's health. It was she, not my father, who faced death. Vivacious Katharine, an uncommonly lovely woman—mother and sister and daughter—was anguished by the wildfire spread of metastatic breast cancer. Katharine's fate had become the family's "extreme reality," as Virginia Woolf once put it.

My father played his role most unexpectedly.

"Rachel," said my mother, shaking my niece's slack shoulder where she lay snoozing in the guest room, after Mum had headed back upstairs for her morning bath. *"Rachel."* My niece opened her eyes, glimpsed an expression of wild vulnerability on my mother's face, and shot into full awakened consciousness.

"Granddaddy won't wake up."

Later that morning, we all received the call, the what-the-hell-are-you-talking-about news that my mother, with Rachel's astonished assistance, dialed out to the family. But Katharine, one hundred miles east of my parents, in Montréal, received her message differently.

"On the night of my father's death," she would tell mourners at his memorial service some weeks later, "I had an extraordinary spiritual experience." My sister, please know, wasn't prone to spiritual experiences. Stress, she was familiar with, as the single mother of two teenaged boys. Laughter, she loved. Fitness of any kind—she was vibrantly physical. Fantastic intellect, fluent in three languages. But she hadn't been paying much attention to God.

"It was about four thirty a.m.," she said, of that night, "and I couldn't sleep, as usual, when all of a sudden I began having this amazing spiritual experience. For the next two hours I felt nothing but joy and healing." There was a quality of light about my sister Katharine, a certain radiance of expression, a melody of voice that hushed every single person in the church—atheist, agnostic, devout. She clutched the podium carefully, determined to be graceful while terminal illness threatened her sense of balance. "I felt hands on my head, and experienced vision after vision of a happy future."

Katharine had described this strange and lovely predawn experience to her elder son as she drove him to high school, before she received the call about Dad. She also wrote about it in her diary: "I thought, is this about people praying for me? And then I thought of Dad cocking his eyebrow, teasing me about hubris." She hadn't known how to interpret the powerful surge of energy and joy she felt in her bedroom—the sense of someone there, the healing hands—until the next day. "I now know that it was my father," she told the mourners. Flat-out, she said this, without the necessary genuflections to science and to reason, no patience for the usual caveats: *Call me crazy, but* . . . "I feel deeply, humbly blessed and loved," she said simply, and sat down.

Astral father, there yet not there. Love flowing unseen. A benign companion of some sort, whose embrace is light but radically moving.

My family is not in the habit of experiencing ghosts. Arriving at my parents' house on March 19, the day after Dad's death, I heard about Katharine's vision for the first time and collapsed to the carpeted floor of the hallway, on the verge of hysterical laughter. My reaction wasn't derisive so much as surrendered. Reality was vibrating, close to shattering.

"Dad is dead, Dad is dead," I had muttered for twenty-four hours already, like a child fervently memorizing new instructions about the way of things, crisscrossing the icy park beside my house, pacing back and forth. *Dad is dead.*

Now Katharine had had a vision.

We took it in as an aftershock. But almost immediately, it began to make profound sense, like puzzle pieces slipping perfectly into place. Without discussing it, we were convinced as a family that he had done something of great emotional elegance. He had died for his daughter. He had seized a mysterious opportunity to go to her, to her bedroom in Montréal, to caress her and calm her before heading on his way.

Later, I learned that this sort of experience when someone has died is startlingly common, not rare, but families shelter their knowledge, keeping it safe and beloved like a delicate heirloom, away from the careless stomp of strangers.

There was much I would learn in the ensuing year about the kept-hidden world all around me, but at the time I understood this much: what a gift this was for Katharine. Waking up over the previous twelve months had meant regaining knowledge of her predicament, which was like an immersive drowning terror in the darkness. How limited we have become, in our euphemistic language, that we speak of patients "battling" cancer, without affording them the Shakespearean enormity of their vulnerability, as if they are pragmatic and detached, marshaling their troops, and nothing like Ophelia, or Lear.

I knew my sister better than anyone else in my life except, perhaps, my children. She was no more or less "brave" than the biblical Jesus when he called out "My God, my God, why have you forsaken me?" She stood keening in the shower the night that the emergency room doctor dispassionately informed her that she had lesions on her brain; she begged an abstracted universe for ten more years to see her sons through school and their own weddings.

Then, suddenly, this astonishment when our father died—*not knowing* that he'd died—of feeling serene, protected, and joyful.

Katharine had watched herself—the future beheld in a mirror, a pool?—as she played with an unborn granddaughter, who she understood to be her teenaged son Graeme's child, on the floor of her bedroom. A five-month-old baby she knew to be named Katie, this wobbly little creature was trying to sit up straight. In her vision, Katharine was holding up the baby girl's back, helping her sit and crushing on her sweetness, admiring the wacky little bow in her hair.

"She was *beautiful*," Katharine told Graeme, of his distant-future fairy child, when she drove him to his Montréal high school that morning. All will be well, and all manner of things will be well.

Back at the house, the phone rang. My mother, calling to report that our father had died.

A month later, in early April, I flew to Arizona to visit the Grand Canyon. A scan had shown that Katharine's cancer had spread to her bones, her liver. "Beauty is only the first touch of terror we can still bear," wrote the poet Rainer Maria Rilke. I read that some time later and thought, *Ah*.

Along the South Rim of the canyon, flies buzzed about the twitching ears of pack mules as they descended the Bright Angel Trail, hooves stepping lively along the steep traverse a few hundred yards ahead of my husband and me. We leaned reflexively into the cliff wall as we followed the party of mule riders, shrinking back from the empty, falling spaces that seemed almost to pull at us, inviting us to swoon and tumble headlong a thousand feet to our doom.

Were only tourists venturing down this crumbling path, I wondered, or pilgrims, too, intent on humbling themselves, feeling their way with hands scrabbling rock, confessing to having no knowledge of the vastness that engulfed them, lured by that very admission? When a scouting party of conquistadors first set eyes on the canyon in the sixteenth century, they chose to believe their eyes, and thought that the Colorado River in the canyon's depth was simply a creek, a thread of blue, easily trudged across at knee's height by their horses. They never did grasp that the canyon was ten miles across, rim to rim, and that the river below coursed as wide as the Nile.

We understand those dimensions now, only because the guidebooks make it plain. We fix our eyes accordingly and, when we spy the river, calculate in relation to what we've been told: if that ribbon down there is actually one mile across, then this height is dizzying. We cower into the Supai sandstone. How do we know what is infinite and what is not? What do we trust—our eyes or our instincts, our guidebook or our gut?

My sister's death was imminent, I felt, perhaps a month away, although she hadn't been given a prognosis—didn't want one—and was still working out at the gym, where people were starting to call her "the Lance Armstrong of Montréal." She certainly looked the part, graceful, agile, and strong. I was acutely aware of her dying, so much so that it seemed to me that the air itself was dangerous to breathe, for each breath demarcated the passage of time. I sensed the clock continuously, how it betrayed me, let go of me, ruined me, and broke my heart with every exhalation.

"It's not that soon, do you think?" my husband asked uncertainly, hiking beside me, aware of my fear of the cell phone in my pocket, of its ring. Well, yes, it was that soon. I'd done the research. Average time to death after brain metastasis from inflammatory breast cancer: three months. But I was alone with this knowledge, because Katharine's oncologist wouldn't say anything out loud that didn't involve metaphors of war. He was currently engaged in bringing out the "heavy artillery," as he called it. Shells were exploding in the rain-dark trenches.

My clock felt increasingly internal and intuitive. When you need to read the world differently, when ordinary channels of information are blocked, what then do you do? About a quarter mile down the Bright Angel Trail, we stopped to rest. My husband went off to make sound recordings, a passion of his, and I crouched in the scant shade of an overhanging rock, perched uneasily on the slope. The view from here was altered, for the canyon now towered above me. My tilted chin faced an immense wall of stone, as tall as a skyscraper. A red-tailed hawk circled high above me in the shimmering air. What I saw, I labeled instantly, unconsciously: a bird of prey, a wall of stone, the quick and apprehensive movement of a ground squirrel. Some tourists, French and German, lumbering along, out of breath, their nylon packs a jarring shade of blue. Someone's dog, trotting; the flies, the mules' dung. And far away a helicopter's burr.

Would a Hualapai woman pausing here a few hundred years ago have broken down this vista into its constituent, material parts? Or would she have seen a landscape rich with portent and spirit, where that bird was not just a bird but a song?

"One has never seen the world well," wrote the metaphysician Gaston Bachelard, "if he has not dreamed what he was seeing."

Father dead and sister dying, time to welcome portent and spirit, even while the doctor yacks on about the efficacy of the latest round of chemo.

We climbed out of the canyon, stopping frequently to take sucks of water from the clear plastic tubes jutting out of our newfangled backpacks. As we approached the rim, I noticed a rainbow. A perfect, vivid little crescent rainbow hanging in the desert sky as if a child had placed a decal on a window. It was so incongruous, given the arid climate, that I chose to make note of it, and checked my watch. Just shy of noon. By evening the sun set breathtakingly, spilling colored light into the canyon. From an Adirondack chair on the porch of the El Tovar Lodge, I called Katharine in Montréal on my cell. No answer.

"Kitty-Kat," I tell her answering machine, "I'm at the rim of the Grand Canyon"—at the end of the world, at the confluence of beauty and terror; here for you, here without you. Katharine, "I'm thinking about you all the time."

She didn't respond, my sister. At the hour I saw the rainbow in the desert sky, just shy of noon, she was being admitted to a hospital in Montréal suffering from acute septicemia, being urged by the doctors to scribble a living will.

A week later, I lay entangled with her on her narrow hospital bed at Montréal's Royal Victoria Hospital, watching CNN on the hanging TV in her curtained-off ward cubicle. In her wisp of a hospital gown, she sipped Pepsi from a straw, bald as an eagle, hands bruised from multiple IV punctures, her legs too pale and slender. My face was tanned from the Arizona sun, while my beautiful sister's was puffed by steroids and flushed from the blood infection that was slowly being brought under control.

She was finally on morphine, and for the first time a little smile played at the corners of her mouth after a weeklong stretch of pained affliction due to wave after wave of intense headaches, with nothing being offered by the hospital but Tylenol because they were treating her for a blood infection and had lost track of the other medical team who had been treating her for cancer. Katharine's characteristic grace and composure masked the degree of her suffering from nurses and doctors on ward rotation, until I had a Tasmanian Devil–style tantrum at three in the morning, when the nurse once again said that Tylenol was the only option "allowed."

So that was where we were, my sister and I, holding hands on her bed and watching coverage of the 2008 U.S. primaries, when her oncologist—finally aware of her existence on this ward being treated for septicemia—came in to break the news that he was transferring her to palliative care. No more chemo. No further radiation. The guns would go silent. It was time, now, he said, to "manage the symptoms."

Katharine moved to hospice on May 14, 2008. The palliative care physician guessed that she had weeks, at the outside margin. But nobody told her that. She was left to envision a horizon without end, distant or near, bright or dark. She didn't ask. Instead, she became a peaceful queen presiding over her court as fifty or more friends, relations, and colleagues arrived for one last conversation, a final kiss. The short hallway of the hospice seemed to me to be streaming to and fro with weeping executives in tony suits, and well-heeled women with red-rimmed eyes carrying bottles of Veuve Clicquot. Just one more toast, another laugh.

The hospice nurses were fascinated, as they told me later, for they were more accustomed to small family groups visiting elderly patients in a quiet, off-and-on way. They watched as we cracked open champagne and played Katharine's favorite songs while she swayed dancingly in her bed, and brought her foods for which she had a fleeting craving, and offered her lilies of the valley to bury her nose in, inhaling. Never have I seen human beings so exquisitely emotionally attuned to one another as we were when we spent those last days in May with my dying sister. When she wanted the volume of energy up, we turned it up. When she wanted it down, we brought it down. The calibration was so precise that when a visitor barged in, all innocence, but with the wrong energy level, we tackled them like a rugby team. Get out, GET OUT!! You're too chipper/too sad/too alpha/too can-do.

When I kissed my sister's cheek she would kiss me back and

behold me in a manner that was so loving it startled me. Generous love, released from need. Often, we sat about wordlessly as she slept, my other two sisters, my brother, and me. Sometimes we massaged her hands with cream and sang softly. Her sweetheart, Joel, played his guitar. My mother, awash in two waves of grief, read Katharine the love poetry that our father had penned for her in the early fifties.

One afternoon, Katharine's ex-partner came by with a vast bouquet of spring flowers that, he explained, had been left anonymously on their former shared doorstep.

"Everybody in the neighborhood loves you, Katharine," he said with a fervent sincerity.

"Surely there's *someone* who doesn't love me," she responded with dry amusement.

She spoke very little, in these final ten days of her life. A few sentences here and there, more often just a word or two. Yet it was clear from everything she said that she was present and observing. Which was why it grew remarkable to us that she seemed so content. She enjoyed our company and the music we played, and gazed admiringly at the garden beyond her window, and the light playing in the curtains.

"Wow, that was strange," she remarked, upon waking up once, her expression one of smiling delight. "I dreamed I was being smooshed in flowers."

All this appeared to interest her—to interest and to please her, as if she were engaged in a novel and pleasant adventure. She looked gorgeous, as if lit from within. Sometimes, she would have happy whispered conversations with a person I couldn't see. At other times, she'd stare at the ceiling of her room as a full panoply of expressions played across her face: puzzled, amused, skeptical, surprised, becalmed, like a spectator angled back in a planetarium, watching a heavenly light show.

I watched her ardently, but she couldn't translate it for me. The sister with whom I'd shared every secret had moved beyond words. "It's so interesting," she began one morning, and then couldn't find the language. "It must be so frustrating," I said quietly, "to not be able to say," and she nodded. We touched foreheads. I was left to guess, or to glimpse what she was experiencing in the accounts of others who'd recovered their voice. I would read later, for example, about the Swiss geologist Albert Heim, who fell off a mountain and wrote, in 1892: "No grief was felt, nor was there paralyzing fright . . . no trace of despair, no pain; but rather calm seriousness, profound acceptance and a dominant mental quickness."

She had not forgotten that she was dying. "Is Mum all right?" Katharine might ask me with concern. Or: "You guys must be falling apart faster than I am."

Indeed, we were. My brain was a computer in meltdown, a car shoved into neutral, an old black-and-white television whose brightness had narrowed to one fizzing star. It is difficult to describe, because I was not capable, intellectually, of observing my own disintegration. I was lost, but Katharine wasn't. She knew very well that she was dying, and more than that. Forty-eight hours before she died, she told us she was on her way. Literally, as in: "I am leaving." How did she know? Hospice could have been two months or six months or two years. If nothing else, hope could have swayed it that way, and she'd subsisted on hope for the first eleven months of her illness. A study conducted by Harvard researchers found that 63 percent of doctors caring for terminally ill patients wildly overestimated how much time their patients had left. The patients themselves, however, can become crisply precise, sometimes nailing their departure to the hour, according to hospice staff.

Katharine woke up one morning and, looking decidedly perplexed, said to Joel, who lay disheveled on the cot beside her: "I don't know how to leave." It was as if she were asking how to water-ski or make bread dough rise. Clearly, she didn't feel anymore the way that we felt, with our thirsting ecstatic joy to find she was still alive each day when we raced to her side. She teased Joel that he looked like a drug addict in his hollow-eyed disarray. She was present, but also elsewhere. Katharine had removed herself to some new plane of consciousness where we were unable to follow.

That afternoon, she gazed through the French doors of her room for a long time with a look that seemed to me, sitting beside her and stroking her hand, to be slightly exasperated. Vexed.

"What are you looking at?" I asked her.

She lifted her arm languidly and pointed in the direction of the garden, remarking: "*Hapless* flight attendants."

We all laughed in surprise. Just then a hospice volunteer wheeled in a trolley of snacks.

Katharine alertly turned to this new visitor and asked: "What's the situation?"

Said the hospice volunteer with brisk cheer: "Well, the situation is that we have lemon tarts, Nanaimo bars, and oatmeal cookies. All home baked."

My sister regarded her as if she were insane.

"I mean," Katharine clarified, clearing her throat, for her lungs were becoming congested, "when do I leave?"

Joel, masterfully suppressing his grief about losing the love of his life after only three years, assumed a comical Indian accent (they'd met in New Delhi) and, wobbling his head, offered: "That is for you and God to decide."

Katharine left the next night, in silence and candlelight, while I lay with my cheek on her chest and my hand on her heart, feeling her breathing slow and subside like the receding waves on an out-going tide. Joel sat on one side of the bed, my sister Anne on the other. The nurse came in, barefoot with a flashlight, to confirm death with a deferential, wordless nod, and we anointed her body in oil and wrapped her in silk. Anne, the only actively religious one of my siblings, offered a Bahai prayer. The staff lit a candle in the hos-pice window. My mother; my eldest sister, Hilary; and Katharine's godmother, Robin, three thousand miles westward in Vancouver, all awoke in their beds, attuned to some new-sounding clock.

"All the world should weep, at the loss of such a lovely girl," Robin found herself dreamily thinking as the curtains shifted and rustled, in the shadows of that dawn.

After Katharine died, I remember my face to the wind, feeling the air, the coolness and fluidity, the urgency. Every day, wind in my face, capturing my attention in some fundamentally new way. I was aware of breath, of what the Greek called pneuma, of a soul-filled world. I gulped the air.

"Welcome to our tribe," someone said to me wryly that summer, speaking of the crazy shift in perspective that comes with grieving. That is exactly how it felt. Suddenly, there were people who under-stood how you could feel like you were gulping air. It was a fiercely intimate bond; even if we had nothing else in common, we had death in common now. It's hard to imagine death and its shared ritu-als being the basis for tribal belonging two hundred or five hundred years ago, but now that the experience of grieving is socially frac-tured, we have no universal consolation to offer, like, "Your father and sister are with God now." Instead, either people feel awkward around you, or those who know all about your wild, unhinged re-configurations say, "Yup, I get it, my friend." Stuff gets weird. You respond to new-sounding clocks, you gulp air.

For a subset of this tribe, the sense that we have encountered a radical mystery unites us as well. We have quietly learned from the dying that additional channels of communication, of which we previously hadn't been aware, enable us to know things in mysterious ways. They enable us to *connect* in mysterious ways—with one another, with the dying, and with the dead—along uncharted or long-forgotten paths.

This shared sense that the dying have opened a door to us that leads elsewhere comes in hushed confidings. During the summer and fall of 2008, people began to tell me things. Some were friends and colleagues I'd known for years, others sitting beside me on an airplane or meeting me for the first time in a bar. If I told them what I'd witnessed with my father and sister, they reciprocated. Almost invariably, they prefaced their remarks by saying, "I've never told anyone this, but . . ." Or, "We've only ever discussed this in our family, but if you think you might do some research . . ." Then they would offer extraordinary stories—about deathbed visions, sensed presences, near-death experiences, sudden intimations of a loved one in danger or dying. They were all smart, skeptical people. I had no idea this subterranean world was all around me.

The director of a large music company drove me home from a dinner party and, when I explained that I was thinking of investigating what my family had gone through, he parked the car outside my house, not ready to say good night. He told me that, as a boy, he had come down to breakfast one morning and seen his father, as always, at the kitchen table. Then his mother broke the news that his father had died in the night. He briefly wondered if she'd gone mad. "He's sitting right there," he said. It was the most baffling and unsettling moment of his life.

On a hot summer afternoon, I stood chatting with a woman on a sidewalk in Pittsburgh; we were waiting for some fellow tourists on a shared weekend trip to the Carnegie and Warhol museums. She told me the story of her sister, who had woken one night to the sensation of glass shattering all over her bed, as if the bedroom window had been blown inward by a tempest. With adrenaline rushing, the sister leaped out of bed immediately and felt around gingerly for the shards of glass she expected to find all over her blanket. There was nothing there. The window was intact; all was quiet. The next day, she learned her daughter had been in a car accident, in which the windshield had shattered. We spoke about another death in her family, and by the time the other tourists rejoined us, there were tears filming her blue eyes. It struck me again how powerful and raw our experiences around death are, how carefully we keep them concealed and yet how close to the surface they stay.

In the late nineties, the palliative care physician Michael Barbato designed a questionnaire for family members of patients after realizing that neither his unit nor most other hospice facilities ever formally investigated experiences like that of my sister on the night of Dad's death. To his surprise, he found that 49 percent of his respondents had had such an uncanny event. "Even if we cannot understand the basis for these phenomena," Barbato argued in a subsequent journal article, "the weight of evidence suggests we cannot continue to ignore them." Certainly, you cannot ignore them when they happen to *you*.

There is pain in loss, and then—in our culture—there is further pain in the silence borne by fear of being dismissed or ridiculed when that loss entails something unexpectedly wondrous. Tell someone your sister felt a presence in her bedroom on the night your father died and, at once, the explanations come:

Hallucination.

Wishful thinking.

Coincidence.

And the implied condemnation: *"Know what?* Yer kinda credulous."

I attended a Christmas party with old university friends, and caught up with a man I hadn't seen for years who works in IT for a bank. I told him about losing Katharine and Dad, some of what had transpired, and he said gently: "I don't mean to be unkind, but it is very likely that she was imagining all these things."

Walking home, I mused about why he found it necessary to say that and why he felt he could speak with complete authority on the subject of what the dying see. He had casually stripped the meaning out of one of the most sacred moments in Katharine's life. Just like that. When I stopped feeling angry, I wondered how he explained away his days. We are meaning-seeking creatures. We dwell among stories and myths; we don't do well chained in all around by a materialist frame and then, for good measure, labeled as fools while we grieve.

"I love you," a man whispers to his new wife at their wedding reception. Imagine a scientist barging into the building: "Prove it!" she commands. "Prove that you love your wife. Do you have the MRI scan?"

Prove your anger, prove your empathy, prove your sense of humor. Nobody really asks you to do that scientifically, of course, because love, anger, empathy, and wit are all considered common elements of human nature, even if they are not scientifically measurable, beyond locating possible neural correlates. Spirituality used to be considered an ordinary part of the human experience as well, but now we call for extraordinary evidence. Why should this be? In some ways it has to do with the rise of what has been called scientism, which is not the use of science as a method of inquiry but rather a prejudice that believes anything that eludes scientific measurement cannot exist.

In 1979, a survey of more than one thousand college professors in the United States found that 55 percent of natural scientists, 66 percent of social scientists, and 77 percent of academics in the humanities believed that some sort of psychic perception was either a fact or a likely possibility. Only 2 percent felt it was outright impossible. In the '80s and '90s, assumptions deepened that all human experience would be explained by the workings of the brain. More recently, some scientists have quietly started to probe that assumption.

In 1999, the psychologist Charles Tart put up a website called the Archives of Scientists' Transcendent Experience, where scientists could anonymously confess their uncanny or spiritual experiences without risk of career blowback. Tart described it as "a safe space" for these scientists, as if they were admitting to a lifetime of boozing or wearing women's underwear. Engineers, chemists, mathematicians, and biologists posted. These scientists, and many like them, put the lie to the persistent belief that only credulous and sentimental people fall prey to certain imaginings. A recent study in the *British Journal of Psychology* showed that there is no difference in critical-thinking skills between people who have uncanny experiences and those who call themselves skeptics. Other studies confirm this lack of difference.

The retired Princeton physicist Freeman Dyson wrote in 2007: "If one believes, as I do, that extrasensory perception exists but is scientifically untestable, one must believe that the scope of science is limited. I put forward, as a working hypothesis, that ESP is real but belongs to a mental universe that is too fluid and evanescent to fit within the rigid protocols of scientific testing."

What he meant was that the tools we have designed to map the genome and determine what makes wheat grow cannot be applied here. The paranormal or spiritual experience comes unbidden. We cannot put my sister in a lab in Palo Alto and wait for my father to die one more time. Dyson received flak for his assertion, but like many of us, he'd witnessed inexplicable phenomena within the confines of his own extended family; his grandmother, he wrote, was a "notorious and successful faith healer." A cousin of his had been the longtime editor of the *Journal for the Society of Psychical Research.*

Skeptics warn that people who engage with the paranormal are either con artists or their credulous victims, and likely a few of them are. But when you know people—when you respect their intelligence, their groundedness, when you witness their discomfort with what they're picking up by unknown means—that characterization is simply unpersuasive. As Dyson said of his cousin and grandmother: "Neither of them was a fool." Nor are the people who have been coming out of the woodwork to tell me of what they'd encountered. Nor was my sister.

Private moments of conversion—from assuming the universe operates by one set of rules to suddenly suspecting there might be other forces at play—can happen to people "like a jolt," as the University of California psychiatrist Elizabeth Lloyd Mayer said, after her own encounter with an act of perception that seemed to be drawing on some unknown sense. In Mayer's case, the jolt came as a result of clairvoyance, which is the ability to somehow glean information across distance. Her daughter had a rare and valuable harp stolen near San Francisco in 1991; neither the police nor the family's public appeals managed to recover it. After several months, a friend suggested to Mayer that she had nothing to lose by consulting a dowser. "Finding lost objects with *forked sticks?*" Mayer responded. But her friend gave her the phone number of the president of the American Society of Dowsers, at the time a man named Harold McCoy who lived in Arkansas.

"I called him that day. Harold picked up the phone—friendly, cheerful, heavy Arkansas accent." She told him she was looking for a stolen harp in Oakland, California, and asked, dubiously, if he could help her locate it. "'Give me a second,' he said. 'I'll tell you if it's still in Oakland.' He paused, then said, 'Well, it's still there. Send me a street map of Oakland and I'll locate that harp for you.'" Mayer sent the map by express post, and two days later, McCoy phoned back to tell her precisely which house in Oakland contained her daughter's harp. Feeling as though she'd surely lost her mind, she put up flyers in a two-block radius of this house, in a neighborhood she was unfamiliar with. She soon received a phone call from someone who had seen the harp in his neighbor's house. He was able to get it back to Mayer. "As I turned into my driveway [with the harp]," she later wrote, "I had the thought: *This changes everything.*"

Mayer needed to completely reexamine her understanding of how the world worked, and after my father and sister died I felt much the same way. I wanted to understand what we knew, and what remained unclear, scientifically elusive, about these controversial modes of awareness. It wasn't enough for me, as a journalist, to accept the officially received wisdom, and it certainly wasn't enough for me as a sister to ignore Katharine's intelligence and discernment and what she was willing to put on the line at our father's memorial service. I wanted to defend her integrity and show respect for our collective experience, so I tried to pursue these questions. Why had my sister had a powerful spiritual experience in the hour of my father's unexpected death? How did she sense a presence in her bedroom, and feel hands cupping her head? Why did she enter into her own dying experience afraid at first—only to become increasingly joyful? What was she seeing, what was she learning, what would she have told me if she could have, after she could no longer converse?

What I learned in the ensuing few years was far richer and more mysterious than I ever imagined, and by sharing it with you, I am hoping that I open a door.

CHAPTER 2

What the Dying See: The Phenomenon of Nearing Death Awareness

The dining room of the West Island Palliative Care Residence is, improbably, a rather cheerful place. When my sister lay dying, it was presided over by a budgie named Blueberry, who hopped from perch to floor to swing to wall and gabbled at all and sundry. On a given day, the sundry includes several volunteers who sing along to lite rock on the radio while they bake scones and whiz up fresh-fruit smoothies for the dying to sip through straws. It has been three years since I ate the delicious soups and tourtières served here to shell-shocked families. Now I am back in a calmer frame of mind, to talk to Monique Séguin, one of Katharine's nurses, about a phenomenon that hospice staff throughout North America call Nearing Death Awareness.

Monique, a middle-aged woman with curling dark hair and an aquiline nose, had seemed bossy to us in our overwrought emotional state, as she shooed a crowd out of Katharine's room one afternoon so that my sister could rest. I even began writing a letter to her at the time, in big, sloppy cursive, that it was none of her business, that no one who was about to sleep forever needed to sleep in the interim. I never gave the letter to her, and I'm glad of that. Hospice nurses and doctors see their patients differently than most families—brand-new to the dying experience—can see their beloveds themselves.

Nurses like Monique have become passionate advocates of creating a hushed, listening space around the dying, because they have learned from experience that the men and women in their hospice beds undergo subtle transformations in awareness and mood. Resting her elbows on the wooden dining room table, Monique tells me that most of the people she's cared for over the years have come to know, at a certain point, exactly when they will die. For the nurses, this certitude is uncanny. "We all know we are going to die . . . one day," said Teresa Dellar, executive director of the residence, in the *Montreal Gazette*. "This is different." Within roughly seventy-two hours of the end of their lives, many dying people in hospice settings begin to speak in metaphors of journey. They are not being *euphemistic*. They are far beyond the task of making everyone feel better. They often haven't said a word in days, and then suddenly they say something focused on travel. They sincerely want to know where their train tickets or hiking shoes or tide charts are.

Monique offers me an example. "We had a patient who was agitated. It was a Friday evening. She keeps saying, 'I want to go shopping.' In life, she was a real shopper. I said, 'When do you want to go shopping?' She said, 'Monday.' I said, 'Fine, let's go shopping Monday.' For me, she was telling me 'I'm going.' And actually, she died that Monday evening."

To families, a desire to go shopping on Monday would have been delusional talk, febrile mutterings of no importance. Far more significant to them, perhaps, would be the anticipated deathbed confession, something for which they had a cinematic sense—a whispered "I love you" or "Take care of the children" before the head falls back onto the pillow. But hospice staff know that when their patients begin to talk about excursions or travel, they are announcing their departure. They do not behave like perishing actors in Hollywood movies. Instead of offering some eleventh-hour contemplation about their lives, they request tickets, or boats. Some ask for their coats, others inquire about the bus schedule. They're caught up in the busy preoccupation of leaving, not reflecting on what they're leaving behind. My sister asked, "When am I leaving?" and expressed frustration about her "hapless flight attendants" in the way I might double-check my flight time to Newark.

David Kessler, former chair of the Hospital Association of Southern California Palliative Care Transitions Committee, has observed this phenomenon countless times—in his own work, and in conversation with medical colleagues. "The notion of the dying preparing for a journey isn't new or unusual," he writes in his book *Visions, Trips, and Crowded Rooms*. "Although, interestingly enough, it's always referring to an earthly journey. People talk about packing their bags or looking for their tickets—they don't mention chariots descending from heaven or traveling to eternity in some other manner." Kessler recalled a ninety-six-year-old man who suddenly woke up in his hospice bed and told his daughter: "Gail, it's time to go."

"Go where?"

"Out! Let's make a run for it. I have to be free."

"She didn't know what to say," reports Kessler. "She helped him sit up, as he seemed to want to get out of bed. 'Is the car ready?' he asked. When she assured him that it was right outside the hospice he said, 'Good. I'm ready, are you?' She asked him where they were going, and he said he wasn't sure. 'I only know that I've got this trip in front of me, and the time has come.' He decided to rest a bit before 'the trip,' and died that morning."

There is no known medical reason for the dying to have such an acute sense of timing about their demise. Palliative-care conferences often devote sessions to how to improve doctors' ability to prognosticate about death. When patients make their announcements about going off on a trip, rarely are there physical signs of imminent decline, such as a marked deterioration in blood pressure or oxygen levels. On the contrary, the bodily symptoms take place afterward. "I'm going away tonight," the blues singer James Brown told his manager on Christmas Day 2006, after being admitted to the hospital for a pneumonia that wasn't considered to be fatal, whereupon his breathing began to slow.

In the most comprehensive, cross-national study of deathbed experiences ever done, the psychologists Karlis Osis and Erlendur Haraldsson confirmed that these intimations of departure even occurred in people who weren't considered by doctors to be terminally unwell. Here is a case reported by one of the physicians to the researchers:

"A male patient in his fifties was going to be discharged on the seventh day after an operation on a fractured hip. The patient was without fever and was not receiving any sedation. Then he developed chest pain and I was called to him. When I came, he told me he was going to die. 'Why do you say so? Having a little pain in the chest does not mean you are going to die.' Then the patient told how immediately after the pain in the chest started he had had a hallucination, but still remained in his full consciousness. He said he felt himself for a few seconds to be not in this world but elsewhere . . . 'I am going,' he said, and departed a few minutes later."

One paramedic sees this puzzling interior knowledge displayed in his ambulance en route to the hospital, as he explained when he called into the radio program *Coast to Coast* with George Noory, during a discussion of the subject. "It's very unnerving," he said. "They know, for whatever reason. They have a prescience. It's a definitive feeling that they have that they are gonna die and I would say ninety-five percent of the time they end up dying in front of me. And it's very disturbing to me." Why does he find it so disturbing? Because it eludes medical logic and thus defies his training. "There were many cases where, really, I did not think that they were ill enough, and then, for whatever reason, they would suddenly have a cardiac arrest and I would say, 'Oh my God, he told me this [was going to happen].'"

Is it a failure of modern medicine to document the mind-body processing of death? "My patient said, 'yeah, I'm going to die today,'" a palliative physician recalled in a 2010 report in the *Canadian Medical Association Journal* about the impact of dying on the personal lives of doctors, but the physician thought, "There is no way that he should die . . . and [yet] he died within forty-eight hours. I marvel at that. There's a mystery there." His reaction, and that of the paramedic's, isn't unusual. "Several medical observers expressed amazement and surprise when confronted with cases in which patients died . . . despite good medical prognoses," reported the two psychologists who compared American and Indian deathbed experiences. "For example, a patient in his sixties was hospitalized because of a bronchial asthmatic condition. His doctor's prognosis predicted a definite recovery. The patient himself expected to live and wished to live. Suddenly he exclaimed, 'Somebody is calling me.' He paused, for it caught him up short, but he tried to dismiss it at first, telling his family, no worries. But, within ten minutes, he had died."

The Florida-based palliative psychologist Kathleen Dowling Singh has described this *way of knowing* as something akin to a transition in modes of consciousness. Sometimes it is fast and so unexpected that there isn't time to understand what has happened. Other times, with a slow terminal illness, it is like a dawning awareness that another realm awaits. She recounts sitting with a nurse colleague who was, herself, now dying and apparently in a coma. Somewhat rhetorically, Singh asked the nurse how she was doing, patting her arm lightly, and surprisingly the woman answered: "I'm halfway there." Writes Singh: "I, of course, will never know whether her words . . . referred to time, i.e. the unfolding of events from Saturday [when she was hospitalized] to Wednesday, when she died, or whether they referred to her psychospiritual movement from tragedy to grace. I do know that she was referring to a process she was aware she was enduring and that that process had, for her, a referential beginning point and end point by which she could measure her halfway point."

Assuming that there is something about dying that is qualitatively different from any previous experience in a person's life, shouldn't that make the process unfamiliar . . . not measurably certain? Ask a woman giving birth to her first child if she knows what to expect moment by moment—if she can tell you with confidence that the baby will come "on Monday." Of course she can't. Why do some of the dying know when they are about to die?

"Does my wife understand about the passport and ticket?"

This question was put—in a hoarse whisper—to Maggie Callanan, a hospice nurse in Virginia, by a well-traveled man who was succumbing to the ravages of pancreatic cancer in the 1980s. She knew what he meant—*does my wife understand that I'm about to die?*— but she couldn't account for the way he was asking, not at this early stage in her career. Callanan first started working with dying patients a scant decade after the hospice movement began to provide families with an alternative to the medicalized hospital death. Trained as an ER nurse, she found that in the quieter realm of homes she could observe a distinct pattern of behavior in her dying patients that didn't correspond to the idiosyncratic nature of drug-induced hallucinations, which she more commonly witnessed in the hospital. But no one seemed to be discussing such behavior in the nursing literature. The pioneering research by Osis and Haraldsson, which they published in a book called *At the Hour of Death* in 1977, had been classified as parapsychology, and didn't find its way into the hands of medical staff. The famed psychiatrist Elisabeth Kübler-Ross had innovated hospice care during this decade and introduced the psychological concept of the "five stages of grief," which was really more about overlapping states of mind—denial, anger, depression, bargaining, and acceptance—that precede the state of consciousness that Callanan was witnessing.

Callanan had read the Atlanta psychiatrist Raymond Moody's groundbreaking research on near-death experiences, which he published in 1975. Certainly, people in palliative care settings can have NDEs days or weeks before they actually die. But what Callanan and her colleague Patricia Kelley were observing didn't really have to do with Moody's classic description of losing consciousness, seeming to drift out of one's body, becoming immersed in a powerful white light, and the other features he described. This was more about what the dying saw, and said, and seemed to feel while still anchored in their full consciousness, in this world. Callanan began to canvass fellow hospice nurses and physicians to find out whether they had noticed these things as well. The uncanny foreknowledge of death's arrival? Visions of deceased friends and relatives in the room? A peaceful radiance? The use of language pertaining to a journey?

Colleagues had, certainly, noticed these patterns but hadn't known what to make of them either. Elisabeth Kübler-Ross was commenting on it in lectures in the 1980s. "Suddenly, someone will bid you farewell when you are not thinking that death will arrive soon," she said, in a speech in Switzerland in 1982. And this puzzled noticing continues: a 2009 study of five hospices and nursing homes confirmed that 62 percent of physicians and nurses had encountered what might be called paranormal "deathbed phenomena" during the study year, but many still considered it taboo to openly discuss what they witnessed.

Callanan and Kelley decided to frame what they observed as a distinct state of consciousness, which they dubbed Nearing Death Awareness. It was critically important to them that families understand what was being communicated by their dying loved ones, rather than dismissing potentially important messages by attributing them to medication or delirium. In 1992, the two nurses published a book called *Final Gifts: Understanding the Special Awareness, Needs, and Communications of the Dying.* The book was like a modern *Arte Moriendi*, a version of the treatises on the art of dying well that once circulated through medieval Europe. Only here, Callanan and her coauthor hoped, nurses and doctors would learn to watch for some of the psychospiritual transformations in dying.

A charismatic Irish American with a keen sense of wit, Callanan continues to see herself as death's best PR person, spreading enlightenment about what the dying process really entails, which she thinks can be far less frightening than most of us tend to believe. "I think we have a moral, ethical, and human responsibility to tell our stories," she told *Vital Signs*, a newsletter of the International Association of Near-Death Studies, "no matter how many times they fall on deaf ears."

Hospice nurse Monique Séguin agrees. "We have to create an opening to be able to listen. If you don't believe in that, well, you're doing your work, as a nurse, but you're missing a few things." Making her rounds at the West Island Palliative Care Residence, Séguin now makes a point of asking her patients about their dreams. It is another way for them to convey how they feel, what they sense coming. They've told her that they dream of riding in a yellow bus, uncertain where their stop is; of floating in a sailboat on a calm pink sea. The woman who had that dream told Séguin, "My [deceased] father was in the boat. My father is coming to get me." Sometimes, they dream of being unable to get their message through to their distraught family. A woman in her eighties dreamed of trying to jam a corncob into a too-small opening, wild with frustration. Whether they speak of journey or through dream logic, Séguin tries to translate for the family when she can.

"I remember we had a patient who kept telling her son, 'Take me home,' and he would just argue with her, 'Mum, you know you're too sick.' She would get more and more frustrated. One evening, I tried to tell him, maybe you should ask her, 'Mum, when do you want to go home?' Maybe she's trying to tell you something. He didn't listen, he was not interested." She smiles and gives a quick shrug; staff, she says, "must tread on eggshells." The woman died a few days later.

One intriguing aspect of Nearing Death Awareness is the tendency for some of the dying to hallucinate visions of deceased family members, friends, and iconic spiritual beings in the days and hours leading to death. Forty-one percent of the dying patients in a study done by the University of Virginia psychologist Emily Williams Kelly reported a "deathbed vision." Are these the kind of sensed presences my sister encountered on the night my father died, there to comfort imperiled, frightened souls? Fifty-four percent of staff in the five-hospice study had patients who experienced a "visit" from a deceased relative very close to the time of death. Informally, nurses often use these visions as a gauge for impending demise. "When a patient says that they have been 'visited' by a dead loved one, you know that their time has come," Penny Sartori, a former critical-care nurse turned academic, told me. She described the first time she encountered this mystery. "When I was a student nurse, I remember crossing over with the night shift, and they said very matter-of-factly, 'So-and-so has been chatting with his dead mother for the last five hours, so he'll be off soon.' I thought they were joking: 'Are they saying that now because I'm new and they are trying to freak me out?' I kept going to check on him, and he *was* talking away to someone I couldn't see. He had a big smile on his face. He died later that day. That spooked me, but I soon realized it was common."

Dianne Arcangel, the former director of the Elisabeth Kübler-Ross Center in Houston, recalls a case in her book *Afterlife Encounters* where she was paying regular visits to an octogenarian weakened by congestive heart failure. He wasn't on death's door, as far as anyone knew, but one day he had his daughter call Arcangel and ask her to come see him. When she arrived at his house, he explained somewhat sheepishly that he'd had a visit from his long-deceased uncle. The relative wanted to reassure him that all would be well, and told him to "ask Dianne what it's like here. She knows." Arcangel was taken aback; she had had a near-death experience several years earlier, and had to assume that was what the apparitional uncle was referring to. Was it? Or did a mysterious hallucination collide accidentally with an enigmatic dream? Whatever was going on, she decided to describe her NDE to the patient, reassuring him that he was destined for a beautiful place.

In Osis and Haraldsson's research, of the 10 percent of the dying who were conscious in the hour before death, the majority reported seeing such visions. Eighty-three percent of the visions were of either deceased people or religious archetypes such as angels (in the United States) and death spirits (in India). It isn't clear what the remaining 17 percent were—gnomes, elephants, pots of chocolate? But the dying have a clear orientation toward the spectral. Sixty-one percent of the patients in this study had received no medical sedation, and 20 percent had been given only weak doses. (It remains rare in India to provide morphine in palliative care.)

"Such experiences can happen to patients who are convinced that they will recover and who are not at all ready to 'go,'" the researchers discovered. "A cardiac patient, a fifty-six-year-old male whose consciousness was clear, saw the apparition of a woman who had come to take him away. He did not seem to be repulsed by [her], just slightly frightened. He said, 'There she is again, she is reaching for me.' He did not particularly want to go, but he did not make a fuss. He became calmer. This experience made him serene. He died a day later."

Searching for other correlations, Osis and Haraldsson determined that less than 10 percent of the patients had high fevers, which can ignite hallucinations. Twelve percent had a "hallucinogenic disease process," such as stroke, brain injury, or uremic disease. But impaired cognition from these diseases "drastically reduced the number of benign apparitions," the researchers found. The more confused or medicated patients were, the *less* likely they were to perceive a consoling or beckoning presence. It is estimated that about half of all Americans who are dying enter (at some point) a state called "terminal restlessness," featuring agitation, anxiety, and flickering psychosis. Some of the factors include organ failure, as well as "opiate toxicity." Here I conjure the memory of my mother-in-law, ravaged by liver and kidney failure in a Nova Scotia hospital, cowering from the onrushing menace of black bears. Her disoriented anguish haunted my husband for months.

These two states of consciousness—Nearing Death Awareness

on the one hand, and terminal restlessness on the other—are radically distinct, for the former is experienced in an oriented and clear state of consciousness, imparts a profound sense of calm, and oftentimes features *information* about the time of departure. As one nurse in a UK hospice study described it, "When they have a high temperature they see things and it's an anxiety-based thing. You can see there's an underlying fear because they don't understand it. . . . Whereas with the end-of-life experience it's like a process and once they have experienced it, they move onto a different level. It's like a journey."

What does that mean? What process? It seems to be an exposure to something, to some state or realm that profoundly—radically— reassures and illuminates the path. In nearly 80 percent of Osis and Haraldsson's cases, the apparent purpose of the deathbed vision was to accompany or take the patient away. They didn't find a single instance of an apparition evincing this "take-away purpose" where the vision involved a still-living person or animal. The bears that menaced my mother-in-law in her toxin-induced psychosis did not turn around and invite her across the River Jordan.

Pondering other angles, the two psychologists examined what they called "the mirage effect." Could these visits from sisters and uncles and angels be the kind of wishful projection that a thirsty desert traveler conjures as a shimmering pool of water? Unable to face self-obliteration, the patient projects a companion to accompany them to their abyss? A couple of factors made that a less straightforward explanation. Patients in distressed or anxious states were less likely to see apparitions than those in calmer moods, their data showed. Also, again, some patients who had the visions hadn't expected to die.

Are families and staff who witness deathbed visions predisposed to see supernatural narratives and coax them along on any pretext? No. In fact, many families are primed in the opposite direction by what our medical culture dictates to them as true. Consider the case of Barbara Cane, who provided this account to the neuroscientist Peter Fenwick: Cane was sitting with her ninety-year-old mother, who had been hospitalized for treatment of pneumonia in December 2005. The lady was lucid, her oxygen levels and blood pressure were stable, according to nurses, and Cane's family was discussing their Christmas plans, among other subjects. At the same time, however, her mother became aware of "these people" who were in the room. She referred to them periodically, as they apparently drew closer and closer to her bed. Not in a menacing, creeping way, but gently and incrementally. Finally, Cane told Fenwick, "she said she wouldn't be there the next day, as 'these people' would 'pick her up when she fell and take her on a journey.' The following day, which was Christmas Eve, the ill woman died. The Cane family hastened to reassure Fenwick: "We think that there is probably a medical reason why the dying hallucinate—poison in the brain, medical drugs, or a lack of certain chemicals in the blood—but it was so strange that it was amidst totally normal conversation, and that she knew so much about her impending death."

What's strange, actually, is that a family would respond to their own observation that their mother wasn't dying alone by feeling compelled to reassure a doctor that there must be a medical explanation.

One summer afternoon, I went to interview Audrey Scott, who was dying of cancer. We hadn't met before, but she had invited me, through mutual friends, to come talk with her. The afternoon was suffocatingly hot, and I found her tangled in a light sheet on a borrowed hospital bed in the middle of her living room, squeezed in between some creaky couches and piled-high tables, her face hidden beneath a cooling washcloth, her body as slight as a bird's. It felt as if the house she had lived in for decades was quietly absorbing her.

Audrey lived in a cluttered bungalow in a tiny town, the property shaded by maple and pine, house cats languishing in the heat as an occasional car rumbled by. Audrey was fading in body—of a cancer metastasized to the bone—but at eighty-three, she remained sharply aware, and was preoccupied, on the day I came, with getting a book she'd written about some early adventures in her life back from the local print shop before she lost consciousness forever. She wanted to approve the final version.

I pulled a stiff wooden chair beside her bed and sat down. She lifted the washcloth slightly to appraise me with keen blue eyes. Her skin was smooth and translucent. The temperature was ninety-five degrees Fahrenheit, and her forehead was sheened with perspiration.

"I'm honored that you've allowed me to come," I told her. We clasped hands, and regarded each other frankly. Dying has a tendency to dissolve all pretension.

We spoke for a few minutes about her book, its manufacturing costs and other pragmatic matters, as if we were meeting in a coffee shop rather than at her deathbed. "I don't want to just give my book to friends and family," she emphasized. "I want it to be a bestseller."

It was, apparently, a collection of her letters home from Europe when she'd been a college graduate making the grand tour, which had sparked a lifelong ardor for architectural history. She'd created a coloring book featuring the Victorian-era buildings of Brantford, Wayne Gretzky's hometown, which was a few miles down the road.

Was this comment about bestsellerdom a genuine lunge for glory at the end of her life or a wry joke? There was no way for me to know. I was learning about who she was just as she took her leave.

"What do you want people to know about dying, Audrey?" I asked.

"There should be no fear," she said without hesitating. She spoke declaratively, with a touch of impatience, hinting at a lifetime of answering the questions of her fourteen natural and adopted children. "Life is laid out from birth to death; it's all just part of the process."

"Are you experiencing . . . or seeing . . . anything unusual?" I wondered, having bugged out the night before trying to figure out how in the hell to word this question.

She had repositioned the washcloth over her eyes, but now eased it back onto her forehead and studied me, a note of caution in her expression. "I see things twirling in the room," she offered. "It's quite pleasant actually." After a pause, she added: "My son Frankie has been visiting me. He sits there." She gestured toward an armchair to my left.

Our mutual friend, Judy, who had been standing discreetly near the window so as not to interrupt the conversation, reached up to the window ledge behind Audrey's head and selected an old, seventies-era varnished frame containing the photo of a smiling young man with thick, square glasses and flat bangs. This was Frankie, a boy that Audrey and her husband had adopted after he'd been disabled in a car accident. He died of cancer in 2002 at the age of thirty-five, Judy later explained.

I angled the frame toward Audrey, so that she could see too, but she evinced no interest, clearly feeling no need for nostalgic glimpses in picture frames if the young man had been sitting right here in the armchair. I tried to find a clear table surface where I could set the picture down, to no avail. I held on to Frankie, uncertainly.

"Is this a dream you are having, or are you awake?" I asked.

She shrugged, determined to remain pragmatic. "I don't think I can tell the difference, with all this morphine."

"Is he talking to you?"

"We've been talking about my books." Audrey's curiosity was so intent that she seemed to be listening even when she was talking. She wasn't seeing or dreaming about anyone else, she said. Not her late husband or any of her living children or friends. No bears or Virgin Marys. For whatever reason, she was encountering Frankie.

I asked her if there was anything she wanted to know about my research, whether she wanted me to describe, for instance, what I knew at that point about near-death experiences. Her attention seemed to quicken, and she nodded. She wasn't much one for "Sunday school stuff," as she put it, but she hoped she was heading to a "place of well-being." Without pain. She studied me from beneath the washcloth. A cat wandered through the room. Cicadas kept up their whirring chorus in the yard.

I told her that others had conveyed an encapsulating and loving light when they'd gone through the brink of death. It felt safe. At some point in this brief description, I had a curious sensation that I was vibrating. It wasn't in my throat or voice, where grief lives, but more in my torso. It had the thrumming feel of an impersonal energy, rather than the riled-up rush of nerves.

I didn't—and still don't—know what to make of the sensation. We frame experiences according to known categories almost instantaneously. *This is nausea, this is anger, this is pain.* If a sensation doesn't immediately make sense to our brain, the nature of it quickly becomes elusive. Was it vibration, or am I just assigning it that description for lack of a better one? Maybe I was suffused with the gravitas of what I was saying, my responsibility to a dying woman who listened to me with hope.

Later, a friend who had been volunteering in hospices described a weird sense of surrounding energy, which had almost caused him

to pass out. So did my sister's best friend, who, while in the West Island Palliative Care Residence, was massaging Katharine's temples
and suddenly realized that if she didn't sit down, hard, on the floor,
she would faint. Maybe it has to do with the life force receding, and
it's like being too close to a whirlpool or a riptide.

"Thank you for that, Patricia," Audrey said. "I'm going to have
a nap now." Judy and I went out into her garden, carrying a box
of her unpublished papers and books, which we sorted through in
the shade. She had written and illustrated a number of small books
about her adopted children. Her drawings were accomplished and
charming, and the accompanying verses were sweet. I realized that
she had devoted a large part of her life to refashioning the life stories of broken, injured people: a blind, mentally disabled child from
India; a girl, now middle-aged, with the mental age of a six-year-old;
Frankie. Had he come back, now, to help her on her way to some
new healing?

Audrey died, at her home, ten days later.

A year after the deaths in my own family, my mother and I flew
to France. We landed in Paris, where Katharine had been born. My
mother wanted to make a kind of pilgrimage to reexperience the tra-
jectory of her daughter's life. Here, on the Rue de Bellechasse, was
the flat where Mum found herself pregnant with her second daughter in the summer of 1955. Here was the little park she wheeled Katharine to in her pram. Here, the narrow road where her "quicksilver
child" had a first tantrum. And here was the school where my sister,
ever charming, won a very French prize for "coquetry."

We took the train to Bordeaux, and then drove to the limestone
caves of mid-southern France where so many famous prehistoric
paintings have been found. One morning the rain poured down as
we scurried into the shelter of a cave and, shaking the wet out of our

hair, looked around. This was a maw, a wide and shallow entrance to a cavern that ran for a mile into the rock body of France. There was a little electric cart to run us along tracks into the lightless interior, led by guides who periodically went insane, we were told, from their mole-like travels underground.

The trolley juddered from the entrance into the darkness, following a rickety track, every now and then whirring to a halt. The guide would disembark with her flashlight and illuminate the close cave walls. A pale beam of light showed that someone had scratched graffiti on this rough limestone during the French Revolution. "Pierre was here," that kind of thing. A shivery spark of time warp. It was another half mile into the black enclosure before a true depth of history began to reveal itself in the dancing flashlight. Cave bears had built nests here, scraped-out hollows of stone, wide as Jacuzzis. Seventeen thousand years ago, humans lay in these hollows and sketched brilliant, perfect art on the ceiling. The guide's flashlight washed back and forth across this ancient overhead and we stared wonderingly. The images above us were utterly commanding. Clean, bold lines—there is no redo option when you're working with charcoal and limestone by rag light. One chance, only, to evoke an aged mastodon with a clear limp, a horse with an evident temper. Perspective, dynamism, a sort of Sistine Chapel for the natural world.

What was their purpose, doing art so profoundly deep in a cave? Had they seen the art at Chauvet in France's Ardèche region, created seventeen thousand years *earlier*? These artists weren't amateurs, and research suggests that their drawings involved an early version of animation, with multiple legs etched in such a way that flickering torchlight would depict fluid movement. These minds were intelligent. Later, we visited France's museum of prehistory and saw an exhibit that argued that "cavemen" did not, in fact, galumph about with shaggy, indifferent hair. They had styled haircuts. Well, of course they did. If they could do art that way, and could carve and play ivory flutes, they could visually conceptualize style.

It hit me then that a great prejudice of our time is to assume that pre-enlightenment humans were stupid. Caught up in a notion of evolution as some sort of linear progress from hunched-over dumbo to straight-backed citizen of the age of reason, we have lost the capacity to believe that the men and women who preceded us could actually have been observant and skeptical, humorous and wise. We assume that spirituality was invented as a hedge against death anxiety by a superstitious populace. The idea that the spiritual world was evident to people—to the dying, to their families, to their shamans—has lost traction. In France's limestone caverns, I began to wonder if that was based on evidence or prejudice. What do we *really know* about spiritual experience?

The science journalist Jeff Warren, who has written extensively about neuroscience and consciousness, told me, "When it comes to the mind, we just do not know whether the brain is a producer or a transmitter. Does the brain generate the mind, like a lamp produces light? Or is it more like a prism or a lens, refracting a preexisting phenomenon into the full spectrum of our personality? Many philosophers have argued that it is at least theoretically possible that 'mind'—the capacity for experience—is some sort of fundamental cosmic property, like space or time. There is nothing about neural activity, per se, that can tell us which of these types of functionality is true."

The earliest scientific investigators, at least for whom we have records, weren't trying to determine the soul's existence—they took that for granted; what interested them was where it was anchored. Where was the "prism or lens"? In the third century BC, the Greek physician Herophilus of Alexandria became the first known man to dissect a human corpse out of pure curiosity. One of his missions was to find the locus of the soul. He decided that it was in the fourth ventricle of the brain, a radical departure from the classic Egyptian understanding, which held that the soul was in the heart. The heart-centric view was the basis on which mummified corpses had their (irrelevant) brains pulled out while their hearts were carefully kept within them to be weighed by the god Anubis.

In the first century AD, the Emperor Hadrian asked Rabbi Joshua ben Hananiah to show him the "soul bone" that some Jewish spiritual authorities claimed existed. Called "the luz," this bone was located somewhere along the spine and was rumored to be indestructible.

"He had one brought," Hadrian later wrote, "and put it in fire, but it was not burnt. He put it in a mill, but it was not ground. He placed it on an anvil and struck it with a hammer. The anvil was broken and the hammer was split, but all this had no effect on the luz."

This account prompted men for several centuries to hunt for the luz, at various times announcing its location in the sacrum, the coccyx, and the sesamoid bones of the big toe.

In the second century, the Roman physician Galen, a doctor to gladiators who peered into their stabbed torsos when he could, (autopsies were forbidden), theorized that the soul animated consciousness in a manner that was mechanically similar to how Roman bathhouses were heated. Humans drew into themselves with their breath the soul-force of the world, and this "spiritus" flowed through pipes and organs and was heated and cooled and performed various functions, such as digestion.

René Descartes decided that the soul resided in (or was somehow received and transmitted via) the pineal gland, which we now understand to govern hormone levels. Descartes believed that spirit flowed to and from the pineal gland through a nervous system that he modeled on a church organ, with microscopic bellows.

In the mid-nineteenth century, an American doctor named Duncan MacDougall decided to try to locate the soul by weighing it. He managed to procure the cooperation of some patients dying of TB in a home for consumptives in Dorchester, Massachusetts. As death approached, he moved his first subject to a cot mounted on a scale and then waited patiently for the last exhaled breath. Whereupon his eyes shot to the needle on his giant scale. The soul, he determined, weighed three-fourths of an ounce. (Hence the Hollywood movie *21 Grams*.)

At around the same time, the very existence of the soul began to be called into question, and the hunt shifted from locating it to proving it could be anywhere at all. These new investigators turned to inferential observations. The nearest analogy may be the methods used in astronomy. Scientists don't actually see black holes through telescopes, but infer their existence by observing that planets and stars are being affected by a massive gravitational pull from certain parts of the universe. What could it be? Likewise, palliative care doctors and researchers observe that something is having an effect on dying people. Whatever is affecting them is inconsistently related to medication, body chemistry, or any given psychological state. What, therefore, is it?

———

The first modern report of a deathbed vision was provided by Lady Florence Barrett, a pioneering feminist obstetrician who was married to a physicist at the Royal College of Science in Dublin. On January 12, 1924, Lady Barrett attended the birth of a child whose mother, Doris, lay dying from complications and blood loss. "Suddenly," Lady Barrett wrote, "she looked eagerly towards one part of the room, a radiant smile illuminating her whole countenance. 'Oh, lovely, lovely,' she said. I asked, 'What is lovely?'

'What I *see*,' she replied in low, intense tones.

'What do you see?'

'Lovely brightness—wonderful beings.' It is difficult to describe the sense of reality conveyed by her intense absorption in the vision. Then—seeming to focus her attention more intently on one place for a moment—she exclaimed, 'Why, it's Father! Oh, he's so glad I'm coming, he is so glad. It would be perfect if only W. [her husband] would come too.' Briefly, Doris reflected to those in the room that she should, perhaps, stay for the baby's sake. But then she said, 'I can't—I can't stay; if you could see what I do, you would know I can't stay.'"

At this point, Doris saw something that confused her: "He has Vida with him," she told Lady Barrett, referring to her sister, whose death three weeks earlier had been kept from her because of her advanced pregnancy. "Vida is with him," she said wonderingly.

Hearing of this experience from his wife, Lady Barrett, Sir William Barrett decided to formally investigate. He solicited written accounts of Doris's apparent vision from his wife, an attendant nurse, the resident medical officer Dr. Phillips, Matron Miriam Castle, and from Doris's mother, Mary Clark of Highbury, all of whom had been in the room. The descriptions corroborated one another, which prompted Sir William, by this time retired from the Royal College of Science, to pursue other cases, which he published as a compendium in a book in 1926 simply titled *Death-Bed Visions*. This book later inspired the psychologists Osis and Haraldsson to compare American and Indian deathbed visions. (Like that of most of the people who come to this research, Osis's own experience originally motivated him to study the dying. As a teenager in Lithuania, he had been with an aunt who had deathbed visions.) Ever since the book's publication, research seems to have been going in dialectical waves: a wave of scholarly curiosity in the early twentieth century; a wave of backlash. A wave of inquiry in the 1960s and '70s; another wave of backlash.

A phenomenon that may be related to Nearing Death Awareness, called "terminal lucidity," originally caught the attention of doctors working in nineteenth-century insane asylums. Patients with severe and chronic mental illness or dementia would suddenly become clear before death. Their psychotic or amnesiac symptoms resolved; they recognized family members for the first time in years; they were able to orient themselves and to say good-bye. Several alienists, as the asylum doctors were then called, made careful note of this in Germany, France, and America. A paper was published on terminal lucidity in the USSR in the 1970s and nurses and physicians began to notice it when hospice treatment became more common in the United States. Elisabeth Kübler-Ross corresponded with Karlis Osis about their shared observations of schizophrenics and stroke patients who suddenly became oriented, direct, and crisp ranging from an hour to a day before they died.

In 2007, the physician Scott Haig wrote an account of his patient David, who had lung cancer spread profoundly and aggressively to the brain. David's speech grew slurred and then he became incoherent. As the cancer cells replaced normal brain tissue, he lost the ability to speak, and ultimately to move. A brain scan done by his oncologist showed that there was scarcely any brain left. "The cerebral machine that talked and wondered, winked and sang, the machine that remembered jokes and birthdays and where the big fish hid on hot days, was nearly gone," wrote Haig, "replaced by lumps of haphazardly growing gray stuff." Lung cancer cells. For days, his patient had "no expression, no response to anything we did to him."

When Haig made evening rounds one Friday, he noticed that David had lapsed into what is known as agonal breathing—the gulping and gasping that accompanies the active dying process. But an hour before David's death, he woke up, and talked calmly and coherently with his wife and three children, smiling and patting their hands, before returning to his dying.

As Haig wrote, "It wasn't David's brain that woke him up to say good-bye that Friday. His brain had already been destroyed." So what was it?

In another example reported by the psychiatrist Russell Noyes, a ninety-one-year-old woman had lost her capacity for speech and movement as a result of two strokes. Yet she suddenly broke through those walls before her death. She smiled excitedly, turned her head, sat up without effort, raised her arms, and called out happily to her deceased husband. Then she lay back down and died. Whether or not she was hallucinating her husband, the far more difficult and astonishing fact for observers to explain was that she'd regained her speech and mobility.

Hospice still serves only a portion of the dying population, although it has been steadily increasing since Maggie Callanan and Patricia Kelley coined the term "Nearing Death Awareness" in 1992. That year, 28 percent of dying Americans were able to pass away in a hospice setting. By 2011, 44.6 percent had. The UK lags behind. Less than 20 percent of British citizens passed away in nonhospital settings in 2008. In Canada, it's about 30 percent. Nevertheless, slowly, the hospice movement is beginning to return families to a forgotten experience of intimate death, and those families, along with attending staff, are starting to challenge twentieth-century assumptions arising out of the machine age. We may owe something *to* the machine age for this shift. Today, pain control has improved. Doctors are beginning to be able to calm distress without wiping out lucidity. The dying may finally be able to convey to us what they are feeling, and where they glimpse themselves to be going.

Who is telling them where they are going? That is the next question, as unsettling as it is. What actually happened on the early spring night when my father died?

CHAPTER 3

Signals and Waves: Uncanny Experiences at the Moment of Death

A humid night in summer, no sounds but the incandescent humming of the streetlamp, the tick of an invisible clock on an antique dresser. Ellie Black rouses quietly to consciousness at around three, her eyes unfocused, mind placid. It's not time to get up yet; nothing is troubling her sleeping child. There is a smell of August grass and last evening's cigarettes. In the stillness, a movement at the end of her bed commands her attention. There, amazingly, she sees her father. Why is he here, she wonders, now fully alert, this difficult man from whom she's so long been estranged? Why *here in her bedroom*? And what on earth is he wearing? Is that a top hat and tails? Her father gazes back at her happily, tips his hat, and bows with a flourish. He is bidding her—his audience?—some sort of farewell. Then he's gone. She blinks. Her bedroom reverts to shadow and silence.

The following morning, she related the experience—whatever it was, a waking dream—to her daughter at the breakfast table. My childhood friend Michele remembers the breakfast conversation with her mother clearly, because she was so surprised when the phone rang later that day, bringing news of her grandfather's demise.

That seeming whisper across the universe, a susurration or hint transmitted by some unknown current the way that birds bend their wings in unison, or ants follow their invisible queen: humans clearly and repeatedly encounter some kind of unexplained attunement. Research done in Wales, Japan, Australia, and the United States shows that between 40 and 53 percent of the bereaved experience "anomalous cognition" when someone close or connected to them has died. Usually, they sense a presence; sometimes they see or hear one. Psychiatrists call these experiences "grief hallucinations," although they have not been studied neurologically. We don't know what to call the intimations—like an estranged father at the foot of the bed, that are our first gleanings of death.

In 1991, the British neurosurgeon J. M. Small wrote to the medical journal the *Lancet* to describe a perplexing experience he had had. "Sir," he began, "what are those waves of communication, that extra sense not yet understood? Something remarkable happened to me." He went on to describe a Sunday morning, "when crossing the hall to the kitchen to make tea, a presentiment of doom beset me and I feared we had been burgled. When I opened the kitchen door all appeared normal, but then there seemed to be a curious descending dark shimmer in the far part of the kitchen, immediately gone—but I knew it was death and female. I thought some catastrophe to one of our daughters-in-law. Disturbed by these suppositions and deciding not to tell my wife, I made the tea and took the tray to the bedroom. As I reached the bedroom, the doorbell rang and I was not surprised to see the village policeman."

One of the two elderly sisters who lived next door to Small had just died in the hospital; the policeman had received the message on his radio, and thought Small or his wife might know the surviving sister and could help him break the news. Small was shocked. Why him? Had the dying sister been trying to recruit his help for the living one? "Was that the cry? My wife and I did have to support the sister, a woman we did not know who had a considerable disability." To the *Lancet*, he concluded: "As a neurosurgeon my mind has been pragmatically directed and I had had no interest in telepathy or extrasensory perception. Here was the reception of information from a source I did not know nor comprehend when it declared its nature, female death. . . . For me to have received such a message remains astonishing. It would be valuable if declared telepathic communicators could be investigated by scanning and electroencephalography to find which areas of the brain are involved with inception, reception, and onward conscious recognition. There was a message in my mind. How it reached there is not defined; although at first confused with fear, it was so very clear."

In 2012, the psychologist Erlendur Haraldsson reported a comprehensive study he had done on 340 cases of extraordinary encounters around dying and death. They happened to men, to women, to young and old, to scientists and sailors, to the bereft and to the content. They happened at night and in the day, waking or napping, traveling, or working. Most commonly people encountered their fathers or mothers, as if the parental impulse to connect and to reassure continues past death. About a quarter of his subjects saw or heard the deceased person either at the hour of death or within the day. In 86 percent of those cases, they weren't aware of the death yet by ordinary means. Thirty-eight percent of the subjects had not expected that such an encounter was *even possible*.

A musician, Rory Magill, sent me a letter that captured this sense of being absolutely mystified and at the same time moved by the symbols and portents of spirit:

"The night my dad died," he wrote, "I dreamed about him passing.

I had last seen him three weeks prior. He had suffered something like a stroke. He was unable to speak or otherwise communicate, apart from some signs of recognition in his eyes, and the touch of his right hand. He was also completely naked, lying under a sheet, for his own comfort, and his scalp was shaved clean on one side for medical purposes. So, he had a striking new look, which remains vivid in my mind's eye today.

"When I left to return to university, the doctor's prognosis was for a fairly good recovery. Three weeks later, I dreamed of my father. He was lying on a hospital bed, on top of the sheets, and he was wearing beautiful new yellow pajamas. Rich yellow, like the color of zucchini flowers. And he had a full head of hair. I was delighted. I climbed onto the bed and held him in my arms, but in an instant I was standing alone in the room, he was gone, and the bed was now empty and neatly made, set in a different corner. The beautiful new yellow pajamas were folded on the pillow.

"I woke up perplexed by the dream but, strange to say now, I didn't make much of it. It didn't trouble me much, oddly, but stayed in my head and colored my day somewhat. When I returned to the rooming house in the late afternoon, I found a handwritten note pinned to the front door. It said, RORY CALL YOUR MOTHER.

"I went immediately back down the street to the nearest phone booth with my heart speeding and my throat beginning to close. I dialed my mum's phone number and my head started to spin. I had no conscious idea of what would come next as the phone rang, but then the instant she picked up at the other end, I started to sob and I just knew. My dad had died while I slept and dreamed of one last visit."

Cross-cultural surveys show that "about half of all spontaneous [telepathic] experiences occur in dreams, and many of them involve accidents or the death of a family member," according to Dean

Radin, a scientist at the Institute of Noetic Sciences in Petaluma, California, who is arguably the most accomplished investigator of psi phenomena in the world today. (Psi, short for *psychic*, refers to cognitive abilities that can't be accounted for through identified senses, including clairvoyance, telepathy, and precognition. Don't look these up on Wikipedia, because at this writing, activist skeptics are editing paranormal topics to present them as having been officially debunked.) Odds against chance in a review of spontaneous telepathy studies have been calculated, Radin says, at "22 billion to 1." Meaning that it could be a coincidence that you had that particular dream on the day when someone you loved died—but it is very, very unlikely.

"I've never experienced that sort of communication so intensely before or since," Rory concluded in his letter, "but it made it abundantly clear to me, for all time, that the signals and waves are transmitting."

———

In the late nineteenth century, the German psychiatrist Hans Berger, creator of the electroencephalogram (EEG), was prompted along the path toward his invention because he wanted to locate the "psychic energy" that somehow enabled his sister to know when he was almost run over by a horse-drawn cannon, a day's distance from where she lived. Picking up on her brother's acute terror in the moment of near collision, she was abruptly beset by an urgent certitude that he was in danger. She begged their father to contact him, refusing to relent until he sent an inquiring telegram. Berger was fascinated by his sister's belief. Do we have a form of consciousness—a way of knowing—that has yet to be charted? he wondered. In 1929, Berger unveiled the first technique for "recording the electrical activity of the human brain from the surface of the head" by using electrodes to detect and transmit electrical signals or brain waves to a graph. Because he was a relative unknown within German medical circles, his invention was initially greeted with skepticism, but his measurement technique was eventually tested by others, and soon adopted around the world, laying the foundation for modern neuroscience.

Three decades later, the EEGs of distance-separated twins were studied, and tentatively found to correlate.

One of the first of the twin studies showed that, when one twin being monitored by EEG was asked to close his or her eyes, which causes the brain's alpha rhythms to increase, the distant twin's alpha rhythms also increased. Twin studies using EEG were subsequently performed more than a dozen times, refining protocols and controlling for design flaws. They continued to confirm subtle correlations in the brains of the separated siblings. In 2013, a study of British twins reported that 60 percent felt they had telepathic exchanges and 11 percent of identicals described themselves as having frequent exchanges with their sibs, including shared dreams. What can science contribute to those claims? After reviewing the methodology of all the EEG experiments done over a decade or so, the Czech neurophysiologist Jiri Wackermann concluded in 2003 that, "We are facing a phenomenon which is neither easy to dismiss as a methodological failure or a technical artifact, nor understood as to its nature." What, indeed, is its nature? How else to get at this question?

In the early 1960s, the University of Virginia psychiatrist Ian Stevenson began to investigate what he referred to as "telepathic impressions." Like Hans Berger, he was interested in determining how people could know that someone emotionally close to them but physically distant was dying or in distress. Unlike Berger, he hadn't had such an experience himself, at least not that we know of, but he was aware that the Rhine Research Center, established by Duke University in North Carolina, had collected and archived thousands of such "spontaneous psi" cases. The Rhine Research Center is generally better known for its early efforts to study telepathy through controlled experiments, like card guessing games. But its researchers have also been interested in these spontaneous cases that arose from the rough, swift-moving emotions of human life. Stevenson wanted to know more about the collection. What did they mean? Could they be verified and analyzed? Would a pattern emerge?

A quiet and meticulous scholar originally from Montréal who would go on to head UVA's Division of Perceptual Studies, Stevenson decided to start by reviewing the first collection of spontaneous telepathy cases known to have been investigated: 165 reports published in the late nineteenth century by the Cambridge scholars Frederic Myers and Edmund Gurney. One of the criteria that Myers and Gurney used for including cases in their book was that some action had been taken *because* of the experience, but *before* the corresponding event was learned of through conventional channels. For instance, a sudden declaration before others that "so-and-so has died!" in advance of the news arriving by cable or messenger, or an insistence on sending an inquiring telegram as Hans Berger's sister had. Myers and Gurney felt such actions were a means to validate the report.

They also included rare cases of collective perception they had come upon: for example, a man and his son simultaneously saw the face of the man's father near the ceiling of their parlor at precisely the time (they later learned) that the father had died. The man's wife, who was sitting in the same room, had witnessed the reactions and comments of her husband and son, although she did not, herself, perceive anything unusual. Another instance of shared perception in the late 19th century, investigated by E. M. Sidgwick, involved distress—a storm at sea—rather than death: "Mr. Wilmot and his sister Miss Wilmot," it was reported, "were on a ship traveling from Liverpool to New York, and for much of the journey they were in a severe storm. More than a week after the storm began, Mrs. Wilmot in Connecticut—worried about the safety of her husband—had an experience while she was awake during the middle of the night, in which she seemed to go to her husband's stateroom on the ship, where she saw him asleep in the lower berth and another man in the upper berth looking at her. She hesitated, kissed her husband, and left.

"The next morning Mr. Wilmot's roommate asked him, apparently somewhat indignantly, about the woman who had come into their room during the night. Miss Wilmot [the sister on board] added her testimony, saying that the next morning, before she had seen her brother, the roommate asked her if she had been in to see Mr. Wilmot during the night, and when she replied no, he said that he had seen a woman come into their room in the middle of the night and go to Mr. Wilmot." This would, of course, have seemed terribly inappropriate in the mid-nineteenth century. *A woman in our room?* Good heavens. Mrs. Wilmot, back in Connecticut, was equally bothered by the impropriety: "I had a very vivid sense all the day of having visited my husband. I felt much disturbed at his [the man in the upper berth's] presence, as he leaned over, looking at us." Still, the experience or dream or whatever it was seems to have moved her. "The impression was so strong that I felt unusually happy and refreshed."

In carefully reviewing such collected curios from nineteenth-century England, Stevenson found that the cases broke down roughly equally between men and women. Eighty-nine percent occurred when the person was awake, rather than dreaming or dozing. (Oliver Sacks steps lively over this research in his recent book, *Hallucina-*

tions, noting simply that "one suspects" the percipients were mostly snoozing. They were not.) Two-thirds of the gathered cases involved news of an immediate family member. Eighty-two percent pertained to death, or a sudden illness or accident. People did not, apparently, pick up on one another's *good* tidings. "Is it that the communication of joy has no survival value for us, while the communication of distress has?" Stevenson wondered. It was impossible to know. In assessing thirty-five of his own contemporary cases, Stevenson discovered that a third involved violent death, whereas only 7.7 percent of all deaths in America in that year, 1966, were violent in nature. (His findings were replicated in 2006, when researchers again found a dramatically higher number of abrupt or violent deaths in telepathic impression cases, in a study in Iceland.) Perhaps, Stevenson mused, there was something in the emotional quality of the event—a thunderclap of fear or pain—that carried like a sound wave across water.

Stevenson was careful with the cases he, himself, chose to research. He excluded "instances of repeated gloomy forebodings which on one occasion happened to be right." He refused to consider accounts without witnesses who could confirm what the original person felt or saw. He interviewed those involved separately, and cross-referenced their descriptions of what happened. In over half the instances, "the percipient's impression drove him or her to take some kind of action apart from merely telling other people about it." A phone call, a frantic trip, an abrupt change of holiday plans. One woman drove fifty miles home in the middle of the night after suddenly gleaning that her teen daughter was in trouble. (It turned out that their house had been broken into by armed intruders, with the daughter inside.) A South Carolinian named Mrs. Hurth provided him with an account, wherein her five-year-old daughter went off to meet her father at a movie theater close to home. Mrs. Hurth saw her on her way and then began doing dishes. "Quite suddenly while I held a plate in my hand an awesome feeling came over me. I dropped the plate. For some unexplainable reason, I knew Joicey had been hit by a car or was going to be." She phoned the theater at once. Her daughter was not seriously injured in the accident, but later wrote her own letter to Stevenson: "I was so terrified (as it happened) . . . I made a silent plea for my mother." Was the child's plea important, Stevenson wondered, in whether her mother caught wind of the cry? Reviewing the cases, he found that it wasn't crucial that "the agent" be focusing on "the percipient" in terms of picking up their signal of distress; but it did affect whether they *took action*. Notably, they responded to cries for help.

How people feel confidence about the telepathic impression they receive is a further mystery. Stevenson found that "a feeling of conviction" was one of the characteristics that separated telepathic impressions from ordinary dreams and anxious imaginings, but it's hard to walk around what that feels like if you haven't experienced it. In a series of email exchanges with a businessman from northeast England, I explored this "feeling of conviction." When he was twelve, Max Bone had a vivid and distinctive nightmare that spurred him to do something that remains unique in his life experience. The dream concerned a house his father had bought, which he was planning to convert into an office. "I awoke in terror," Bone told me, "in the early hours of one morning, after having what seemed like a nightmare, but the content of which had a noticeably different and unusual quality. It had been very windy that night, and I had dreamt that I was outside the rear of this property in Borough Road, standing on the pavement, and facing directly towards the rear yard gate. This green-painted wooden-paneled yard gate off the street was unfastened, and was opening, then banging shut again, and again, in the wind. I approached the gate and the gate opened, the peeling paint and grain of the grey and denatured timber was shown in incredible detail. As the gate swung open revealing the small backyard, I noticed that the white half-panel kitchen door had been pushed completely open, leaving just a dark rectangular hole in the wall. The top pane of the small kitchen window was also broken, and partly open. As I moved through the open gate, and across the yard towards the empty door, fear started to build in me. My vision centered on the small kitchen step. As I reached the threshold I was looking right down at the step, and the fear became incredible, and as I began to cross the threshold I woke up in terror.

"The next morning (I believe it was a Sunday morning), I still recalled the dream in detail, and I did something unusual—I acted on the dream (I have never acted on a dream before or since this event). I immediately went downstairs and told Dad about the dream whilst he was having his breakfast. I explained the dream's unusual quality, and made clear that I believed that something had happened to the property, pestering him to drive over to it immediately.

"Dad finished his breakfast, and decided to indulge me, as he hadn't checked on the house for a while anyway. My two elder brothers overhearing the story, and I think sensing something exciting, wanted to come along too. So my dad, myself and my brothers drove over to the house. We pulled up at the rear of the property, opposite the green gate, which was indeed blowing open and shut in the wind. When we all saw this, both my brothers turned to look at me in the car, and pulled 'spooky' faces at me.

"We left the car and entered the backyard through the gate, to find the kitchen door wide open; it had been pushed right back against the kitchen units so that it was not visible, just like in my dream. The top pane of the kitchen window was also broken and slightly open, again just as in my dream. We entered the house to find water gushing through the ground floor ceilings, and the beginnings of mold on the dining room carpet.

"It became clear that the house had been broken into some days before, and the thieves had been returning to remove fixtures, fittings, and lead piping over some period of time. We secured the house and turned off the water. Dad refused to talk about the incident for many years although he does talk about it now. My two brothers found the whole affair creepy and unsettling and recall the incident to this day.

"I have thought deeply about the incident for the whole of my life, and drawn the best conclusions I can to explain [it.]" Bone keeps an eye on theories and speculations in paranormal and neurological research, and is convinced that there is something in the electromagnetic field that enables thoughts and perceptions to travel between minds. He doesn't think he actually went out of body in his dream. He suspects that he saw, somehow, what the *thieves* saw, tapping into their perception of the back of the house. He doesn't accept that the dream was a coincidence. I pressed him about why he couldn't accept it as coincidence, and he said it was the distinctive *nature* of the dream.

"My dreams tend to be sort of 'softer,' still plenty of visual detail but they don't have the hard-edged detail of this dream. I had absolutely no problem in recalling the dream and remembering I needed to do something about it. That's not typical in my experience. It was almost as if the memory had been lodged somewhere it shouldn't really be lodged. Almost as if 'raw' unprocessed imagery had been laid down, accidentally bypassing my normal visual pre-processing, having what I considered to be almost silly levels of recorded detail, like the peeling paint on the gate." As the sole inhabitants of our heads, we are generally the best judges of a strikingly different perception; it is, invariably, the clarity and specificity of the impression that prompts people to act.

Ian Stevenson found that there were two other factors that made people sit up, wide-eyed, and reach for the phone or the pen or otherwise take action. One was if the "agent," which is to say the person in the crisis, specifically focused on the "percipient" during the moment of danger. This seemed particularly true of parents responding to children, although that would make sense because children would be most likely to cry out to an absent parent. The second factor was, possibly, a higher giftedness for picking up such signals in the first place; a number of his cases involved people who had formed "telepathic impressions" at key moments more than once. Janey Acker Hurth, for example, who had sensed her daughter's imminent collision with a car had also, some years earlier, twigged to her father's sudden illness. She and her husband were newly married and visiting her in-laws when she woke to "a feeling of deep sadness, an impression that something was wrong." She waited for the inexplicable sentiment to fade away, but instead it intensified and she began to sob. This woke her husband, who questioned and tried to console her, to no avail. In the morning, she went downstairs exhausted and bereft. "I put bread into the toaster and while waiting for it I suddenly wheeled around and exclaimed, 'It's my father! Something is terribly wrong with my father!'" A phone call to the house within moments of this exclamation confirmed her sense of how she was narrowing in on the matter. Her father, her mother said on the phone, had fallen into a coma after his kidneys failed in response to a sulfa drug. (He would die shortly after.)

Stevenson was struck by how information sometimes gradually came into focus for people. "The percipient's mind," he mused, "may scan the environment for danger to his (or her) loved ones and, when this is detected, 'tune in' and bring more details to the surface of consciousness."

Stevenson's findings from the 1960s are echoed in cases collected three decades later by the British neuropsychiatrist Dr. Peter Fenwick, of King's College, London, a longtime specialist in epilepsy who has an abiding curiosity about unusual perceptions and experiences around death, sparked when one of his patients described a near-death experience. Fenwick, by now in his late seventies, has amassed more than two thousand accounts of what he calls "deathbed coincidences." Working with solicited written letters from the British public, Fenwick's research is less exacting and more exploratory than Stevenson's. But the accounts give a rounded flavor of what people encounter.

One woman wrote to Fenwick about the husband from whom she'd recently separated, who committed suicide in February 1989.

She awoke at 3 a.m. from an intensely vivid dream in which her ex was sitting on the bed, assuring her that it was over, that he had found peace. "I got up, 'on automatic,' did some work I needed to do, two clients phoned me around eight o'clock and I freaked them out completely as I told them I would be taking some time out because my husband had just died." She didn't yet objectively know this to be true, but *she knew it was true.* She went over to his flat with Merlin, their dog, and she discovered the body. A coroner's report fixed the time of death at around 3:00 a.m. Here, the "agent" in distress didn't seek help by focusing on the "percipient," but rather sent a message of reassurance after all was done. He was dead.

Richard Bufton, a college lecturer and commercial diver, was aboard ship when he learned of his grandfather's death via unconventional means. He wrote to Fenwick: "I was lying in the bunk in the forward cabin in a sort of half-asleep state, when what I can only describe as a vision similar to seeing a teletype ribbon went past my eyes. The words, which I read in my mind, simply said, 'your grandfather is dead.'" Fueled by that feeling of conviction, he dashed into the common area to put through a radio call to his mother in England. "When she answered the phone she said she had some bad news. I interrupted to tell her the reason I had rung was that I knew my grandfather was dead."

What's interesting about Bufton's experience is that, according to the neurologist Dominic ffytche of King's College, London, one of the world's leading experts in visual hallucinations, when people hallucinate text, they don't see meaningful messages. The hallucination is visually incoherent, either a rough approximation of text or a random assemblage of letters. So, if Bufton was hallucinating, how did he see words stating clearly that his grandfather was dead? In the one case that ffytche has found where a woman was actually visualizing written-out "command hallucinations," suggesting that she should throw tea in a family member's face, for instance, he discovered that she wasn't actually reading the words, in the sense of visually scanning each word. Instead, she was inferring them. And that seems to have been the case with Bufton. It was as if he were picking up a message about his grandfather in some other way, and then imaginatively projecting it as a teletype ribbon. Some researchers propose that people intuit these death and distress events, garnering the raw information, and then their brains instantly assemble a representation of what they intuited, much the way we impose meaning in our dreams on external sounds like a ringing alarm clock.

When someone appears at the end of the bed, are they reaching out to you, postmortem, or are you perhaps sharing telepathically in their dying experience of calm and peace? Impossible to know.

Several people have reported the strange joy that my sister Katharine felt, as if she received from my father not news of his death but a shared sense of his final elation. Sudden peace, buoyancy, contentment, or alternatively sorrow or physical pain. A sailor named Raymond Hunter appears to have shared both the pain of his father's illness and the peace that followed. Of the evening his father died of lung cancer, he said, "I cannot possibly describe the feelings of love and peace I experienced." But these emotions flowed after a much more bizarre and intense interlude where he felt as if his lungs were collapsing and he could scarcely choke in breath. This abrupt and violent experience of another's dying symptoms has been noted by some researchers, although it remains almost completely unexplored. Ian Stevenson came across it in his cases, and suggested that it could be a kind of telepathic extension of a more commonly documented phenomenon in which people who live together sympathetically take on one another's symptoms or moods. "The syndrome of couvade, in which a man imitates the symptoms of his wife's labor pains and delivery, has been well documented with numerous examples," Stevenson noted. He, himself, had had a patient suffering from right shoulder pain with no identifiable cause, whose son had just died of a cancer that generated right shoulder pain. "Pain due to identification and mediated normally through the senses raises the question," he wrote, "whether a similar identification can take place by means of extrasensory perception."

A particularly vivid instance comes from an interview conducted by the journalist Paul Hawker in 2010. A woman in her late thirties told him:

"I was awoken around 2:00 a.m. by the sound of my heart breaking. I know that sounds really odd, but that's what it was. I heard it crack and felt my chest sort of splitting. It was massive, sudden and explosive. The next morning I did all my usual early morning things and got into the car to drive to work. I was sitting at a set of the traffic lights when I became aware of or felt this pressure on the side of my face. I distinctly remember that the pressure was that of a cheek lightly pressed against mine, sort of cuddling me. The feeling I was filled with at this time was one of love and support—it felt fine. I then felt a hand holding my hand and 'felt' it had no middle finger. I knew this because there was no pressure in this area. And then it dawned on me. I realized it was my dad's hand; he'd lost his middle finger in a building site accident when I was a little girl.

"I continued on to my first appointment, which was a short one, and I returned home after an hour to be met by my husband's words, 'Your dad's gone.' Apparently he'd died from a massive heart attack during the night. I wasn't at all surprised."

———·—·———

Dismissing these experiences as wishful imaginings of events after the fact ignores their intensity, clarity, and power to unsettle. They take place on a different plane than the cool realization that you've just been thinking of someone who then calls on the phone. Consider the sailor Raymond Hunter's description: "I remember grabbing my mouth, forcing it open to help me breathe. I was fighting for all I was worth but the pains were now unbearable." *Unbearable.* That is not something you shake off as a strange bit of dreaming, particularly when you learn that your father died in that moment of your panic and pain.

The sociologist Glennys Howarth of Plymouth University, researching rare cases of shared illness symptoms across distance, observes that this sort of event becomes a kind of crisis of identity, as you pass from one mode of being in the world to another. "If the person sharing such experiences is to make sense of them, for him or herself and for others, a plausible explanation is crucial." ('Someone tell me what the heck happened!') Without an explanation, the person will be stranded "in a stigmatized explanatory world of hallucination and madness." It's not a comfortable place to be.

At the age of eighteen, a man named Derek Whitehead told Fenwick he had been working in the Merchant Navy and was in his ship bunk one night reading a magazine. "I looked up and my grandfather stood next to me, looking at me. Well, I shot off the bed, I did scream, and he was still there looking at me. I ran for my life up to the bridge shaking like a leaf." Whitehead would later learn by post awaiting him in Australia that his grandfather had died that night. He found this highly alarming. "I don't know what these things are—fantasies, dreams, wishes, delusions—I don't like them. They make my sense of reality wobble."

This sense of destabilized reality is, of course, one of the reasons that people sometimes fiercely resist the idea of anomalous experiences. A particularly arresting story comes from the Harvard-educated surgeon Allan Hamilton, a professor of neurosurgery at the University of Arizona Health Sciences Center, who had the following experience during his years as a medical resident in Boston. In 1982, he was doing a rotation at the pediatric burn unit of the Massachusetts General Hospital. A ten-year-old boy was brought in after falling from a high-voltage tower onto a power line. The only skin that remained on his body were patches in the folds of his joints, and at his groin. "In the initial phases of critical burn care," writes Hamilton in a memoir called *The Scalpel and the Soul*, "the victim must be covered with new skin. This is first accomplished with grafts taken from fresh cadavers. Although the skin is dead, the thin strips of dermis and epithelium work beautifully as temporary skin. Soon the patient's immune system rejects the foreign grafts. The hope is that the cadaver grafts will buy enough time that the remaining pieces of the patient's own skin . . . can be gradually harvested to resurface the body."

In this case, the boy's fragile frame rejected the skin grafts continuously, so that Hamilton and his colleagues were losing hope they could keep him alive long enough to regenerate his skin. He remained in a coma, precarious, barely there. His forty-two-year-old father, wildly distressed, collapsed and died of a heart attack. Suddenly, the medical team had an almost imponderable opportunity. Genetically, his father's skin would be a closer match than that of any of the previous cadavers, and there was a chance that the boy wouldn't reject it.

"The decision was made to take Thomas to the operating room and cover him with his father's skin," Hamilton recalls. The grafting took nearly twelve hours. After the operation, "I went into the call room and fell asleep instantly," Hamilton writes. "I had been on the move for more than forty-eight hours straight. Only seconds seemed to pass before I woke up angry and disoriented. A nurse was knocking loudly on the call room door. I looked at my watch. I'd been asleep for over two hours. The nurse was hammering, and it suddenly flashed into my mind that Thomas was probably dying."

Instead, Hamilton found his young patient—who would indeed survive—roused from his long coma, and scrabbling madly with mittened hands at the endotracheal tube in his windpipe. He wanted to talk. Hamilton removed various bits and pieces of equipment from the boy's trachea and mouth. "He coughed violently a couple of times. Suddenly, he spoke. His voice was perfectly clear.

"'What happened to my father?'"

Stunned, Hamilton's impulse was to lie, and he assured the child that nothing had happened, that his father was fine. The boy found this answer confusing: "My dad's just standing there at the end of my bed. Why doesn't he say something?"

Hamilton craned his neck to look for an image or silhouette beyond the hospital curtain that might be tricking his patient, but there was nothing there. He and the attending nurse broke the news to the boy of his father's death, as if it were more important to disabuse him of disconcerting illusions than to protect him from sorrowful news. If the boy was shocked, the doctor was bowled over.

He writes: "Here was my own fragile moment of awakening. It left me tingling all over, as if sparks were dancing off my skin."

Fragile awakenings, private startlements, moments of utter confoundment, for my Irish and Scottish Highland ancestors, the reality of an extraordinary way of knowing things was always there, embedded comfortably within their culture, a deeply ancient experience. One summer afternoon, my elder aunts and cousins, women in their eighties and nineties, gathered around the dining table at our cottage of Lochend on Ontario's Stony Lake, a gathering spot for the Mackenzie clan on my mother's side since we went into exile from the original Lochend near Inverness, Scotland, during the Highland Clearances. Here, in a hundred-year-old cabin in the North American forest, my grandmother had painted a saying directly on the wall: "Fra ghosties and ghoulies and long-leggedy beasties, and things that go bump in the night: the guid lord deliver us." A playful nod to the Gaelic the clan once spoke and our witchy Celtic ancestresses. Now, we had come to talk of such things seriously for the first time over lunch.

Aunt Bea, who is ninety-two, still drives herself four hours to spend summers on an island at the lake, reads the *New York Times*, and has a thing or two to say about British foreign policy and climate change. She keeps her gray hair in a bob with bangs and gazes at you intently from beneath the fringe, always curious and probing. She recalls that Great-Grandmother Maude always went about with an absolute confidence in her mysterious way of knowing. "Granny would be sitting in the living room reading a book or something, and she'd suddenly slam it down and mutter, 'Damn! So-and-so is coming and I don't want to see them.' And sure enough," Aunt Bea says, her arms clasped lightly across her belly, "so-and-so would show up ten minutes later."

The Norwegians have a word for this uncanny anticipation of visitors: *vardøger*. We call it "second sight," which was the term our Highland ancestors used. (As an interesting aside, many Highlanders have been found to have Norwegian DNA, and these are the only two European cultures I'm aware of with folk names for this particular precognitive trait.) Culturally, it was understood that someone could possess second sight; nobody who had the talent felt like a fantasist or a liar. When tiny, slender Great-Granny Maude intuited the approach of some bore or crank at the cottage, she sometimes hid in the painted wooden chest on the porch. Years later, when my grandfather, her son, telephoned her to report his father's fatal heart attack on his sailboat in Lake Huron's Georgian Bay, Maude replied impatiently and disconsolately: "I *know*."

After Maude, with each successive generation, talk of "second sight" subsided as the family merged into the rising secularism of the twentieth century. But at our reunion at the lake house, Aunt Bea, our eldest relation, spoke of Great-Granny Maude, and Cousin Marion offered her own story. She had been working at a resort hotel in Banff, Alberta, as a teenager in the late forties, when the hotel caught on fire, prompting her mother in Montréal to wake in high distress and make an urgent call to her. Cousin Sonny confided that she had drifted into the mystic after an allergic reaction to penicillin—and that later, on the very afternoon when her ex-husband died, she was suffused through and through with a warm, hard-to-describe "glow" of emotion. My mother, the über-rationalist, conceded that, now that we were on the topic, *come to mention it*, she did awake suddenly one morning in university and hurry to the dormer telephone to call my grandmother, whom she knew to be in crisis. Granny's dearest friend had died that night.

Each experience was different, but all were ways of knowing—or modes of being—that tilted the world on its axis, if only for a moment. Yet, we had never shared them before. It was amazing, in a way. Here we had all been walking past the Gaelic incantation on our own wall—fra' ghosties and ghoulies—and completely ignoring that lore's relation to our own experiences. We had long since internalized the masculine, left-brained Scottish enlightenment philosophy of our patriarchs, most of whom (since immigrating to North America) have been soldiers, Anglican clergymen, and accountants. Ghosties, ghoulies, and witchy Celtic women: be gone with you!

What goes missing when we silence the conversation about our impressions and visions? We no longer have the old models for understanding them. "Models help us think," writes Berkeley psychiatrist Elizabeth Lloyd Mayer, who was bowled over by a dowser finding her daughter's harp: "Without a conceptual home, observations that don't fit our existing models may be intriguing and entertaining, but they have the ultimate impact of writing on water. Without a model to contain them, we have no place to put new and unfamiliar things while we try to figure them out."

History is littered with examples of mainstream science deliberately overlooking "new and unfamiliar" things. It's worth pausing to consider the record. The French Academy of Sciences in the eighteenth century scoffed at meteorites because: how could rocks fall from the air? Museum curators across Europe jettisoned the meteorites they had in their collections, as a result, embarrassed that they could have been seduced by something so fanciful. In the late nineteenth century, Hungarian obstetrician Ignaz Semmelweis demonstrated that, if doctors washed their hands before delivering babies, the rates of infection in mothers went down, but his proposition was deemed absurd and he was ridiculed into obscurity, eventually dying, unhinged, in an insane asylum. John Snow was belittled for proposing the existence of germs.

My favorite example is the one that novelist Hilary Mantel pointed out a few years ago in the *London Review of Books*: "From 1904, the Wright brothers made flights over fields bordered by a main highway and railway line in Ohio; but though hundreds of people saw them in the air, the local press failed to publish reports because they didn't believe the witnesses, and didn't send their own witnesses because it couldn't be true. Two years after their first flight, *Scientific American* dismissed the feats of the flying brothers; if there had been anything in it, the journal said, would the local press not have picked it up?"

About ten years after Ian Stevenson began investigating telepathic impressions at times of crisis, three psychologists developed a method of measuring telepathy in the lab that came to be known as the "ganzfeld technique." *Ganzfeld*, roughly translated from the German, means "whole field"; the premise is that you can widen your field of perception and tap into subtler sources of information by quieting everyday sensory input. It is not dissimilar in principle to hushing a noisy group of hikers in the woods so you can listen for and hear the snapping of a twig by some small, concealed animal. "The ordinary waking state is largely driven by sensory awareness, so anything that disrupts that awareness will probably improve psi perception," Dean Radin writes.

In a typical ganzfeld experiment, the person being tested as the receiver settles in a comfortable chair with halved Ping-Pong balls over their eyes. Their eyes remain open, and a red light is shone at their face, so that their visual perception is of a mellow red glow. They wear headphones, so that their only auditory stimulus is the rhythmic whooshing of unpatterned sound. In this calm, reduced-stimulus state, they are left to settle for a quarter hour or so.

Meanwhile, in a separate room, the volunteer playing the role of "sender" opens an envelope containing four distinct images. Choosing one, he or she attempts to mentally project the image to the receiver, for a period of thirty minutes. In an experiment run by Radin in his lab at the Institute of Noetic Sciences in Petaluma in 2010, the pictures that a participant named Tom could choose to project included a grassy field with blue and yellow flowers, a bird's nest with four golden eggs, the Great Pyramid of Cheops, and a plain asphalt road flanked by telephone poles in a flat landscape.

The receiver, Gail, was told to vocalize her impressions without naming or analyzing them. She was recorded as saying:

Keep feeling like looking up at tall. I'm looking up at tall.

Something about texture. Texture.

I feel like something has a rough texture.

Tall, very tall impression, looking up high.

Feel as if I'm walking around observing something, like when you would walk in an art gallery or in a museum and you would look at something.

Wow.

First I'm feeling like tall trees, and then I'm feeling like a tall building.

And then I'm like a Yosemite kind of image of a tall rock or a tall, some kind of a very tall solid stone something.

Seeing browns and grays.

Something like a feeling of walking around, looking up and being in awe.

In awe of something.

Monolithic, or I don't know what the word is.

Of the four pictures he'd found in his envelope, Tom had opted to mentally project to Gail the one depicting the pyramid of Cheops.

Between 1974 and 2004, nearly ninety ganzfeld experiments were conducted and published by a number of scientists around the world. Overall, randomly selected pictures were described correctly 32 percent of the time, which is 7 percent above what you'd expect to see by chance. In other words, the odds that someone like Gail would describe an image approximating the picture Tom chose *by chance* and do so 32 percent of the time have been calculated at 29 million trillion to 1. It can be difficult to get one's head around these kinds of statistics. Seven percent above chance doesn't sound like a lot. But it's enormous: it shouldn't be happening above chance *at all*. If telepathy is a gift for attunement, like musical genius or mathematical intelligence, and you're sampling a random number of university volunteers for your experiments, none of whom is picking up on resonant, echoing distress or death calls from their families—just mundane stuff like pictures of an Egyptian pyramid—then 7 percent above chance is extraordinary.

Radin's books take readers step by step through the history of the research, the refinements of protocols, the selection of sub-jects, the "file-drawer effects," and the calculation of odds. The Brit-ish psychologist Richard Wiseman, a popular skeptic, has said that "by the standards of any other science," certain psi phenomena, like clairvoyance, have been "proven." But he argues, paranormal phe-nomena are extraordinary and should be held to higher standards of evidence. In 1988, the National Research Council of the National Academy of Sciences commissioned the Harvard psychologist Rob-ert Rosenthal to scour the research for methodological flaws. After sifting through the studies, Rosenthal and a colleague reported that "the ganzfeld ESP studies regularly meet the basic requirements of sound experimental design." Furthermore, they reported to the NRC, it would be "implausible" to say that these telepathy findings resulted from chance. According to Rosenthal in an article later pub-lished in *Psychological Bulletin*, the NRC Committee Chair responded to this news about telepathy by asking Rosenthal and his colleague to withdraw their findings. They refused.

"It is a scandal," said the Cambridge scholar Henry Sidgwick, "that the dispute as to the reality of these phenomena should still be going on, that so many competent witnesses should have declared their belief in them, that so many others should be profoundly inter-ested in having the question determined, and yet that the educated world, as a body, should still be simply in the attitude of incredulity." Sidgwick made that protest—one imagines a waved walking stick and a mustachioed scowl—in 1882.

One hundred and twenty years later, much progress has been made in studying these signals and waves, but what is considered official truth has not changed. The Cambridge physicist and Nobel Prize winner Brian D. Josephson told the *New York Times* in 2003: "There's really strong pressure not to allow these things [psi phenomena] to be talked about in a positive way." Harold Puthoff, a physicist at the Stanford Research Institute appointed to oversee the CIA's remote viewing (or clairvoyant) experiments in the 1970s and '80s, described this pressure in a series of emails to the psychiatrist Elizabeth Lloyd Mayer. "The evidence we had [a clairvoyance] was rock hard," he wrote to her. "I saw that. But I also saw that it didn't eradicate my doubt. That made me see my doubts weren't the problem. On the contrary, the problem lay with my beliefs. I was having terrible trouble giving up my beliefs about how the world worked, even in the face of evidence that said my beliefs were wrong."

By the time of the first EEG twin studies, physicists had discovered the quantum universe, where subatomic particles were breaking all the rules of classical physics and demonstrating a phenomenon called "entanglement." Two particles with no physical link to each other could somehow remain connected or entangled, exerting influences on each other at a distance. This is, by now, well-established and discussed. Physicists have grown to accept "nonlocal" connectivity as a real, if totally mystifying, phenomenon at the subatomic level. In 2011, physicists at Oxford took quantum entanglement to the macro level by briefly entangling two separated diamond crystals that were visible to the eye.

The neuroscientist Michael Persinger, of Laurentian University in Sudbury, Ontario, thinks that he may have demonstrated the entanglement effect between people, although his experiment needs to be replicated. "What we have found," he reported in 2009, "is that if you place two different people at a distance and put a circular magnetic field around both and you make sure they are connected to the same computer so they get the same stimulation, then if you flash a light in one person's eye, the person in the other room . . . will show changes in their brain as if they saw the flash of light."

Scientists like Persinger (who used to be an adamant skeptic about all things paranormal) are increasingly at ease with the findings in this new world. "Quantum theory and a vast body of supporting experiments tell us that *something unaccounted for is connecting otherwise isolated objects*," writes Radin. "And this is precisely what psi experiences and experiments are telling us. The parallels are so striking that it suggests that psi is—literally—the human experience of quantum interconnectedness."

Scientists based at Princeton are exploring this mysterious force of connectedness as well, through what they call the Global Consciousness Project. Beginning in the early nineties, faculty at the Princeton Engineering Anomalies Research lab began placing electronic "Random Event Generators" at collaborating universities around the world. REGs are essentially like coin tossers: they generate heads and tails, or ones and zeros, following nothing but the laws of chance. Preliminary lab experiments, however, had determined that REGs could begin to behave *less randomly* if they were the focus of the researcher's willed intention. Pursuing this idea that human consciousness could somehow exert an effect on material systems, lead investigator Roger Nelson and his team began monitoring data during events of global significance, such as the 1997 funeral of Princess Diana, which is estimated to have been watched by 2.5 billion people, and later, the calamitous unfolding of the attacks on the World Trade Center of September 11, 2001.

"We asked if groups of people brought by circumstances into resonance or coherence might share a group consciousness that would register in the data from our random devices. The answer was yes," writes Nelson. For seventy-two hours, beginning on the morning of 9/11, the REGs were feeding patterned data from all around the world into the Princeton lab. You can see what this looks like, how random data shifts to patterned data on the PEAR Global Consciousness Project website. It makes for an eerie kind of art. They have even converted the data into sound, creating, for instance, a dirge for 9/11.

"The overall statistics for the project," Nelson wrote, "indicate odds of about 1 in 20 million that the correlation of our data with global events is merely a chance fluctuation. And we can exclude mundane explanations such as electromagnetic radiation, excessive strain on the power grid, or mobile phone use . . . We don't yet know how to explain the correlations between events of importance to humans and the GCP data, but they are quite clear. They suggest something akin to the image held in almost all cultures of a unity or oneness, an interconnection that is fundamental to life."

It is a unity glimpsed in biology. In 1919, the American naturalist William Long published a book called *How Animals Talk*, reporting his observations of wolves in the Nova Scotia wilderness. The pack he was studying appeared to be able to range beyond the threshold of hearing and smell, and yet still keep track of one another. Long found this interesting, although he wasn't in a position to prove anything, one way or another, about what it meant. It would be another sixty years before the maverick former Cambridge biologist Rupert Sheldrake picked up on Long's field observations and applied it to a study of dogs.

Sheldrake designed a series of experiments to test how dogs know—or seem to know—that their owners are coming home. Controlling for scent, the sound of the car on the road, the routine time of day, and all familiar sounds, Sheldrake was able to establish that dogs begin to anticipate their owners' arrival regardless of the sensory cues. A force appears to bind and alert social animals over distance, Sheldrake concluded. Traditional human cultures take this for granted, although it hasn't been widely studied by anthropologists. (Who's going to give you a research grant to study telepathy in indigenous tribes when it isn't supposed to exist?) Intriguing hints appear here and there in field reports. The South African journalist and author Laurens van der Post described African bushmen who knew when a hunting party would return with a kill; they explained to him that it was like the white man's "wire"—or telegraph—but, they gestured at their chests, "in the heart." The Iroquois of North America refer to such communications as using "the long body," which denotes the means by which they stay connected to the group, to their tribal lands, and to objects on that land.

Reading this, I realized with a start that I'd encountered similar stories myself, in the summer of 2010 when I flew more than a thousand miles north of Toronto to visit the Ojibwa and Cree on their traditional lands in the boreal forest near Hudson's Bay. The Elders there were born in the forest, following an age-old way of life. They had scarcely more contact with modernity than the tribes of the Amazon or the Andaman Islands, until their children were brutally forced into boarding schools in the twentieth century, as late as the 1970s. These Elders spoke of their shamanic "shaking tent" ceremony, which they had used, for all intents and purposes, as a radio. The shaking tent consists of long strips of hide forming a narrow sphere, almost like a closed umbrella. The shaman stands inside it, although its width is barely larger than his body. As he enters into a trance, the hide strips begin to flutter, as if he's created some kind of energy field. But, explained the Elders, the shaking tent isn't merely a ritual means to commune with the spirits. It was how they learned what was going on before they had telephones—what the weather was like, what hunting conditions other families and groups were encountering hundreds of kilometers up and down their main waterway, the great Severn River. Like the Masai with their "telegraph," there was—there always had been—this other way to know things.

Standing in the midst of an unutterably huge spruce and pine forest, I lose any armchair skepticism and grow humbly curious. How did human beings evolve and maintain a culture and economy over such inconceivable distances, patrolled by bear, wolf, and wolverine? One starts to appreciate the evolutionary advantage of developing an alternate means of perceiving or communicating, a sixth sense, that could transcend distance limitations.

Factoring in the number of Ian Stevenson's telepathic impression cases that were akin to distress calls, rather than simply intimations of death, the evolutionary advantage becomes all the more clear. "It is altogether probable that important unrecognized exchanges of feelings through extrasensory processes are occurring all the time," Stevenson concluded, and "even if we can only observe it occasionally, and usually between persons united by love and during a special crisis to one of them, this should arouse our curiosity and our efforts to find out why this is so—why the union is latent for most of us and why it does sometimes reach expression in a few of us."

Dean Radin offers one theory about what might be going on to facilitate our gleanings of danger. "At a level of reality deeper than the ordinary senses can grasp our brains and minds are in intimate communion with the universe," he has written. It is as if we inhabit a sort of matrix of mutual awareness, like a pond which feels everywhere and nowhere the ripple of rain. It extends, he suspects, beyond the ordinary boundaries of space-time, connecting us all at once with future and past, near and distant shore. "From this perspective, psychic experiences are reframed not as mysterious 'powers of the mind,' but as momentary glimpses of the entangled fabric of reality." He goes on: "Particles that are quantum entangled do not imply that signals pass between them. Entanglement means that separated systems are *correlated*."

If you grow alert to the fact that someone you cherish is in danger, "it would appear to be a form of information transfer, but in fact it would be a pure correlation. That is, within a holistic medium we are *always connected*."

We are always in a position to feel what others feel, and to learn of their joy or distress, even to react to that joy or distress without knowing why, but most of the time we're not paying attention. We can't; if we did, we'd be overwhelmed by the signals and waves. It would be like straining to hear your name whispered in a nightclub at two in the morning. For survival's sake, the brain has evolved to filter *out* most information. So, even when the connection flares, we may be unclear what we're responding to.

"When we were students at Bristol in the late 1960s," a woman wrote to the neuroscientist Peter Fenwick in London, "my fiancé was estranged from his mother, who was a doctor in Nigeria, and he was living with his father. Around New Year, we were both attending a dinner party at my parents' house when he suddenly began to cry uncontrollably and was consumed with grief. He went out to the kitchen to wash dishes, trying to give himself something to divert his mind, but nothing worked, and eventually he gave up and drove back to his father's house, where he was met by a policeman who told him that his mother had been killed that evening . . . he was not at all an emotional man, quite the opposite, and this strange episode has remained as one of the most inexplicable episodes of my life."

In 2010, the Cornell psychologist Daryl Bem published research in the rigorously peer-reviewed *Journal of Personality and Social Psy-*

chology establishing that people were able to unconsciously intimate events a few seconds in the future. Recruiting more than a thousand student volunteers, Bem conducted nine separate experiments on well-established psychological effects, such as reacting to subliminal images or arousing images, but he time-reversed them. For example, in one experiment, the volunteers were told that they were going to be shown two pictures of curtains side by side on a computer screen. They were told that one curtain had a picture behind it and the other had a blank wall. Thinking they were being tested for ESP, they were asked to choose the curtain that they felt had the picture behind it, and that curtain would then draw back to reveal whether they were correct. In fact, there were no pictures behind either curtain. Once the volunteers had made their choice and clicked, the computer randomly assigned a picture—neutral or erotic—to a random curtain. And here's what happened: the volunteers clicked on erotically arousing, rather than neutral, pictures at an above-chance rate, and *before the pictures were there*. In other words, they reacted to the reveal of an erotic image by clicking on it in advance of it actually being there to arouse them.

The nine experiments were, according to colleagues, impeccably done, notwithstanding the fact that no one could believe the results were true. As one of his peer reviewers, Joachim Krueger of Brown University, put it: "My personal view is that this is ridiculous and can't be true. Going after the methodology and the experimental design is the first line of attack. But frankly, I didn't see anything. Everything seemed to be in good order."

Bem tested for personality traits in those who performed particularly well on his experiments. Two things caught his eye. One was that the better performers scored high in the trait known as extraversion. Such people tend to be stimulus seekers, restlessly scanning the environment. Maybe, he speculated, they were more apt to pick up on remote or obscure signals. The other element Bem noted was that people who did well with precognition tests also tended to be very quick at processing subliminal data. Both of these traits would, Bem theorized, have had a considerable evolutionary advantage, all the more so if people could pick up on cues across time and space.

"The ability to anticipate and thereby to avoid danger confers an obvious evolutionary advantage that would be greatly enhanced by the ability to anticipate danger precognitively," he wrote. "It was this reasoning that motivated (our) experiment on the precognitive avoidance of negative stimuli. Similarly, the possibility of an evolved precognitive ability to anticipate sexual opportunities motivated (our) experiment on the precognitive detection of erotic stimuli."

Bem was building on the insights of the psychologist Hans Eysenck, who argued that "psi might be a primitive form of perception antedating cortical developments in the course of evolution." If so, he wrote, the later evolution of "cortical arousal might suppress psi functioning. Because extroverts have a lower level of cortical arousal than introverts, that provides another reason (besides enhanced stimulus-seeking tendencies) for predicting that they will perform well in psi tasks." All evolutionary psychology, even when it includes parapsychology, is purely speculative, bear in mind.

Experiments in presentiment have also been done recently by measuring the nervous system response to stimuli—a scary face, for instance. The face might trigger a physiological reaction, such as changes in skin temperature (blushing or blanching) or accelerated heart rate. But in these experiments, such responses were triggered a few seconds prior to the scary face *actually appearing*. Research done by the Spanish biologist Fernando Alvarez in Seville has found that

Bengalese finches show alarm up to nine seconds before the video monitors next to their cages actually display a horseshoe whip snake seeming to approach them. Other experiments in presentiment continue apace all over the world, studying (in no particular order) college students, earthworms, zebra finches, and Zen meditators.

In 2012, the Northwestern University neuroscientist Julia Mossbridge and colleagues published an article in the journal *Frontiers in Perception Science* that concluded that carefully sifted studies dating back to 1978 established a small but statistically significant incidence of precognition in experiments without methodological flaw. Before agreeing to publish Mossbridge's article, Mossbridge told me, one of the peer reviewers for the journal requested that a line be inserted to say that this phenomenon was due to natural physical processes, if not yet determined. Yet, says Mossbridge, there is no reason to assume that presentiment is anything other than a natural physical process, simply a process we don't understand.

One hopes that this kind of research will one day explain the premonitions and presentiments that people receive almost daily—and in every era. Abraham Lincoln dreamed of his death three days before Booth shot him and wrote a letter about it. During World War II, prisoner of war Lev Mishchenko in Stalin's Gulag in 1949 dreamed of his lover Sveta in a white dress, kneeling by the side of a little girl. After he was released, in 1962, Lev and Sveta were walking to the lake across a field that skirted the forest. Lev was in front, Sveta behind him with their daughter Anastasia, who was then six. "As I reached the edge of the forest," Lev recalled, "I had this feeling . . . I turned around and behind me I saw Sveta in a white dress kneeling on the ground to adjust something on Nastia's dress. It was exactly what I had seen in my dream—Sveta on the right and, on the left, our little girl."

Does this explain my sister, in the quiet of her bedroom, having a vision of her future unborn grandchild?

Somehow, the universe connects us and consoles us. The dying reassure us—and *are* reassured. Sometimes, as I explore next, these presences we witness and dream of actually, even, assist us.

CHAPTER 4

Astral Father: The Phenomenon of a Sensed Presence

War is a theater for ghosts. Filled with corpses and fraught with danger, the frontlines are nightmare places with moments of unexpected grace where the signals and waves are transmitting like mad. There are many accounts from the two World Wars of families receiving the "three a.m. news"—a telepathic impression—that their boys had been felled. This accords with the research by Stevenson and later by Haraldsson that violent and accidental deaths result in more telepathic impressions. But intriguing tales also abound of soldiers themselves sensing a guiding presence alongside them in battle. Those who are dying or in peril seem, then, to encounter these presences regardless of whether they are lying quietly in bed or crouching in a rat-infested funk hole. The "grief hallucination" is, at the same time, a guardian angel; the "deathbed vision" also a guide.

In his memoir about fighting in World War I, the journalist William Bird described what was, for him, the most extraordinary chapter in the battle. He was in France in April 1917, sleeping beneath a groundsheet amid the muddy and mazelike trenches after the Battle of Vimy Ridge. In the cool darkness, he was awoken by the grip of someone shaking him. He tried to pull away, exhausted and irritable, but the grip held with some urgency, so he opened his eyes and saw, to his confusion and surprise, his brother Steve, who had been reported missing in action two years earlier.

"Steve grinned as he released my hands, then put his warm hand over my mouth as I started to shout my happiness. He pointed to the sleepers in the bivvy and to my rifle and equipment. 'Get your gear,' he said softly."

While Bird tried to work out in his mind how his brother had even located him, he obediently followed Steve away from the other sleeping men and down the trench. By the time it occurred to him to ask where they were going, his brother rounded a corner—and vanished. Bird searched for him frantically, but eventually reconciled himself to the conclusion that he had been asleep on his feet, dreaming. His brother, come to think on it, had been kitted out in the uniform and cap worn in 1915, two years out of date. It had to have been a dream for sure. In despair, Bird gave up looking for him, and fell asleep where he was, after crawling into a funk hole.

The next morning, he was awakened by his battalion mates, who were excited to have found him alive. They took him to the bivvy where he'd been sleeping before he'd moved to show him that it had been hit by a high-explosive shell. The bodies of the men who had remained there were beyond recognition. The incident was such a profound part of his war experience that Bird titled his subsequent memoir *Warm Hands*.

"The supernatural and the uncanny seemed natural and explainable in these sites of mass death," writes the military historian Tim Cook. No soldier dismissed his own experience as a product of sleep deprivation or madness, for there was often an element of being assisted or rescued. George Maxwell would write, for example, of being separated from his platoon in no-man's-land on a lightless evening, lost and uncertain of the contours of the enemy lines. Terrified, he was about to bolt in a random direction like a maddened horse when a voice commanded him, "Be seated and await deliverance." Brought up short by the authority of this unseen speaker, he obeyed, kneeling in a crater, which is where one of his comrades located him and led him to safety.

This experience of the sensed presence was also encountered in the extreme environments that Europeans and Americans were first exploring at the turn of the last century—in particular in the Arctic and Antarctic. "Who is the third who walks always beside you?" wrote the poet T. S. Eliot in 1922. "When I count, there are only you and I together, / But when I look ahead up the white road / There is always another one walking beside you." These lines in Eliot's "The Waste Land" allude to the uncanny experience of Sir Ernest Shackleton, who made a desperate, exhausted trek across a mountainous stretch of Antarctica in 1916, climbing and staggering for forty kilometers with two of his crew members in search of rescue after their boat got mired in ice and provisions ran low. At some point in the arduous and frozen journey, having left the remaining crew behind with the stranded boat, all three men became aware of a presence—another companion—accompanying and guiding them. The presence, or to Eliot, the "third man," although in reality it was a fourth, seemed to escort them safely to a whaling station, and then departed. None spoke about it during the trek itself, each thinking that they, alone, sensed the extra companion. Later, when Shackleton was asked about this, he refused to surrender such a sublime experience to ridicule: "None of us care to speak about that," he said. It was, he had come to think, an experience too transcendent to become the subject of casual Ouija Board chatter.

His reticence has been shared by a great many explorers, sailors, divers, and mountaineers who have experienced the Third Man in the midst of duress and danger. Their companions have sometimes been visible, sometimes not; sometimes the presence has spoken aloud to them, other times not. But always the presence has comforted them and, in some cases, it has led them to safety.

Early scientific explanations tended to focus on the location. Soldiers in World War I were presumed to have been sleep deprived. Mountaineers were thought to be suffering from the effects of altitude, lack of oxygen, and cold stress. Victims of shipwreck, who had plenty of heat and air, were assumed to have hallucinated due to sunstroke and dehydration. Polar explorers had been tricked into the illusion by monotony and sensory deprivation—a world of white—in which the brain conjures figures—or ghostly suggestions of figures—almost as a way to keep itself stimulated. (In fact, as has been demonstrated with the ganzfeld telepathy experiments, sensory deprivation may be an ideal environment for enhancing receptivity to subtle information.)

Meanwhile, men traversing the rich and varied terrains of jungles and forests were also encountering the Third Man. When Henry Stoker (cousin to Bram) and two fellow British sailors escaped a Turkish prison during World War I, wandering for days through the wilderness, the Third Man stuck by them until they found rescue. "We had all three been sensible of his presence throughout the most trying part of the night; we all three agreed that the moment he left us was when we felt we had put the danger behind. I cannot exaggerate," Stoker wrote, "how real his presence was, how content one felt—despite the mystery of it—that he should be there."

Three of the World Trade Center survivors on 9/11 later claimed that they were guided to safety by sensed presences. One was prodded through a wall of fire he would otherwise have shied from, fearful of the flames, and led down the stairs of the North Tower; another was comforted as he lay beneath the rubble; a third trapped beneath concrete received encouraging visits from a presence she perceived, for some reason, to be a monk.

In 1989, two mountain climbers, Lou and Ingrid Whittaker, experienced the *same* visual hallucination of a kind, middle-aged Tibetan woman staying with them on India's Mount Kanchenjunga. Lou Whittaker was leading an American expedition when he became aware of the woman, "a friendly spirit," at base camp. She kept him company each evening for three months. Whittaker's wife, Ingrid, meanwhile, was trekking down from the summit to the base camp with her own expedition group when she developed altitude sickness. For several days, she lay in her husband's tent and found herself being attended during the day by a Tibetan woman. "She was wearing a headscarf and a long dress. She was shadowy and two-dimensional, like a silhouette." It was "very comforting." Both of them experienced this woman but made no mention to each other of the vision until months later.

Over the last decade, Swiss neuroscientists have tried to replicate the sensed presence in their labs by using electrodes to stimulate a part of the brain called the left temporoparietal junction. Lead researcher Olaf Blanke, at the Mind-Brain Institute in Lausanne, implicated this area of the brain because it "integrates sensory input into a cohesive picture." For Blanke, any dysfunction in this brain region, due, for example, to "disruption of the oxygen supply," might cause a person to get disoriented as to self and other, or here versus there.

His research is pioneering, but unfortunately there isn't any evidence yet that people encountering the Third Man are all experiencing a disruption in the normal functioning of this part of the brain.

They might be, for reasons as yet unknown, but so far no scientist has managed to place mountain climbers and lost soldiers into laboratories to test the theory that disturbances to the temporoparietal junction (TPJ) are creating these experiences out in the quixotic world. "Explanations of the sensed presence phenomenon abound, which is paradoxical, given the paucity of systematic research on the subject," notes the clinical psychologist Peter Suedfeld, a leading expert on human cognition in extreme environments. But, "How these hypotheses explain the repeated and/or prolonged appearance of a helpful 'other,' is not clear."

The Third Man is nothing if not helpful.

In his fascinating compilation of accounts, *The Third Man Factor*, the journalist John Geiger is perhaps the first to have explored this uncharted realm of human experience, establishing how extraordinarily prevalent the Third Man is across different landscapes and predicaments. Geiger first noticed the phenomenon when writing about historic polar expeditions. When he extended his gaze, he found dozens of references in memoirs, diaries, letters, and in his own interviews with explorers and adventurers. Nobody, individually, realized how widely shared the experience was, which accounts for why the scientific theorists were acting like the proverbial blind man and the elephant, chalking things up to oxygen deprivation here and sunstroke there.

The journalist Maria Coffey describes a wider range of mysterious and transcendent experiences by endurance athletes in *Explorers of the Infinite*. Coffey—as surprised as Geiger by what she found—encountered a number of typically hardheaded pragmatists who had had precognitive dreams about their coming struggles, telepathic access to distress messages from partners, and sensed, guiding companions. She wondered if these phenomena were common to this

group of people because of the intense focus and heightened awareness they entered into as athletes, which might enable them to tune
into fields of perception that we are ordinarily oblivious to.

What is a sensed presence? What is the nature of the experience, whether in the mountains, in no-man's-land, in the rubble of
earthquakes or broken towers, or in the dark and quiet of my sister Katharine's bedroom? It isn't the shivery sensation you get that
you're being stared at, or a momentary apprehension at the stirrings
of shadow and light. It isn't a wistful belief, like thinking someone's
there when the wind whispers through a candle. There is nothing
speculative about the experience. Those who encounter the Third
Man describe having a sudden, vivid, and indisputable awareness
that someone is with them, sometimes for hours and even for days.

Nor is the presence indifferently with them, the way pop culture
depicts forlorn and sulking ghosts who pass us by in haunted houses.
These sensed presences are *relational*: they are purposefully and sup-
portively with the people who sense them, which is what the psychologist Peter Suedfeld, one of the few academic experts on sensed
presence in extreme environments, points to as the most difficult
part to explain.

Sometimes, the presence merely acts as a quiet friend; Geiger
reports on cases where a climber has offered the sensed presence a
snack before realizing no one is there, or has divided their dinner ra-
tion in two. On occasion, people pause on the trail waiting for the
presence to catch up before remembering with a start that no one
is physically there. Other times, the presence actively guides people out of danger. In 1983 the American climber James Sevigny was
so severely injured by an avalanche in Banff National Park that he
could barely move. His back was broken, as were his arms, his nose,
and some ribs. He was bleeding internally. He lost consciousness for

an hour. When he attempted to stand, he collapsed. He supposed he would die, and was yielding to shock and hypothermia when a sensed presence materialized. "It was something I couldn't see, but it was a physical presence," he told Geiger.

The presence more or less bullied Sevigny to get up and move, prodding him through deep-crusted snow every raw inch of the one and a half kilometers back to his camp. "All decisions made," he said, "were made by the presence. I was merely taking instructions." As soon as he reached his tent, his bossy companion disappeared. Moments later, Sevigny was found by some cross-country skiers and helicoptered to a hospital.

Joshua Slocum, the first man to sail solo around the world, reported that, in the midst of a battering storm, he had fallen ill with food poisoning and couldn't keep to the helm of his boat. To his astonishment, a tall man appeared, and said that he would handle the tiller so that Slocum could recover. Slocum had the impression that this man was "a friend and a seaman of vast experience." The man kept the vessel on course for 145 kilometers, until Slocum could resume control, and then dematerialized.

Several other sailors have witnessed this startling turn of events in moments of danger, finding themselves assisted by a mysterious other when their own strengths were failing; so have pilots. Edith Stearns, a contemporary of the more famous female flier Amelia Earhart, came to expect a sensed presence on her juddering flights in the 1930s and '40s, as if she'd conjured an imaginary friend: "I never fly alone," she told a journalist for *Life* magazine. "Some 'presence' sits beside me, my copilot as I have come to think of it." On one occasion, the presence actively warned her, shouting, "No! No, Edie, don't!" when she began attempting an emergency landing on railway tracks, as yet unaware that a safe airfield runway lay a few miles ahead.

These sensed presences meddle. The pilot Brian Shoemaker, disoriented in his H-34 helicopter during an Antarctic storm, was accompanied by a presence who told him to "turn 20 degrees to the right." He obeyed, because, he admits, "I had nothing else to go by." That steering adjustment got him safely out of trouble.

The presence seems to vanish moments after the danger is resolved, even if that resolution isn't yet clear to the person at risk, which is one of the most striking mysteries about the more long-lingering hallucinations. Wrote the shipwreck survivor Ensio Tiira, who had been adrift on a raft for thirty days, "I'd lost all sense of a second person being in the raft. The guardian angel who kept me company . . . left the raft with my own sense of hope." But the day the angel vanished turned out, in fact, to be the day of his rescue.

The same steadfast accompaniment and sudden vanishing act happened to an American climber named Rob Taylor, who broke his ankle on Mount Kilimanjaro in 1978 and was left to wait at the base of the mountain when his climbing partner struck off in search of medical assistance. Taylor's envisioned overnight stay propped against a boulder turned into a much longer ordeal. He ran out of water and his ankle became badly infected. On the third day, he became aware of a man sitting nearby on a boulder. He assumed, at first, that the figure was a member of the rescue party, but when calls and then yells failed to elicit a response, he grew angry—throwing rocks at the figure—to no avail. He settled into being entirely mystified. At length, he concluded that the man was simply there to keep him company.

"Hour upon hour this companion watcher, as I call him, peers out at me through the curtain of snow," he later wrote. As Taylor's health deteriorated, the figure quietly drew closer until, after a few days, it was "right at the foot of the sleeping bag." Then this benevolent and reassuring presence, whom Taylor reported "took up absolutely solid space like a stone or anything else," suddenly left him. Minutes later, a rescue group arrived.

For Taylor, as for Shackleton and many others who have encountered such a protective presence in times of peril, the world's subsequent cheerful dismissal proved to be painful: "I don't often talk about my companion watcher these days," Taylor wrote. "He is a creature out of place here, misunderstood. After the (rescue), when I first spoke of him to people, they reacted quite predictably: 'What an imagination!' 'Your fever had you hallucinating.' At first I persisted in my stand: 'He was real. There in the flesh or at least in some concrete form I could see.' Later, I left him out altogether. It was easier than trying to define or defend him to people who could not understand. Now I know this and say this to you: He was there and as real as you or I. I do not know to this day his purpose, but I sense that it was good."

Taylor was making a plea for respect for a phenomenon that is best known and keenly observed by the people who actually encounter it. It puts me in mind of a poem by the American poet Mary Oliver about the dying visions of William Blake: "When a man says he hears angels singing / he hears angels singing. / *When a man says he hears angels singing / he hears angels singing.*" It's impossible to objectively measure subjective perception. In the study of consciousness, subjective perceptions are sometimes referred to as qualia. "Something beyond our understanding occurs in the genesis of qualia," the neurologist Oliver Sacks has written; it is "the transformation of an objective cerebral computation to a subjective experience."

What is the quality of redness? How do we invest red with a different feeling than yellow? What is lovable and meaningful about a red rose? What makes the scent of jasmine transporting? What is beautiful about a snowy field? Why are we awed by a sunset? Why is the Third Man so comforting and so real? *We are not in a position to answer objectively.*

"There is a fundamental explanatory gap between brain activation and conscious experience," the Brazilian psychologist Alexander Morelos said at a University of Arizona conference on the science of consciousness that I attended to get a better handle on the subject. The "explanatory gap" is widely referred to among scholars who study consciousness. Most academics are respectful of this gap, but those who are impatient to stride over it tend to engage in what their critics, like Morelos, call "promissory materialism." To wit, they take it on faith that everything we experience is generated by the brain. They figure that sooner or later, the brain will reveal its secrets. We might not understand yet why brains project the image of a consoling, guiding presence like the Third Man, but we can assume the brain is doing it somehow. "But that is a hope," Morelos argued, "it is not a scientific fact."

Part of the problem with that hope, interestingly, is that it has a slightly fantastical quality about it. As the psychologist Julio Peres asked at the same conference: "What kind of empirical evidence would we need to prove that 'the brain believes' or 'the brain interprets'? These are enchanted metaphors. How does it make sense to assign a psychological trait to an organic brain part? What materialists are actually engaging in is animism. This represents a return to a much less critical and more naïve metaphysics than what they were hoping to overturn. We are not explaining anything. We have not progressed beyond the *assumption* that all will be explained by the brain."

In the meantime, neuroscientists are trying to identify if not the causes of qualia then the neural correlates. This refers to the parts of the brain that turn on or "light up" when they are viewed on scans performing certain activities. Brain regions *correlate* with different activities like listening to music or processing a visual image. The music makes waves outside the brain, which are received and interpreted via particular neural pathways. What's so challenging about qualia is understanding why the brain imbues certain bits of visual or auditory or olfactory information with emotional significance.

Wrote Sacks: "Philosophers argue endlessly over how these transformations occur, and whether we will ever be capable of understanding them. Neuroscientists, by and large, are content for the moment to accept that they do occur," and to seek out correlates. For several years, now, the neurophysiologist Richard J. Davidson at the University of Wisconsin-Madison, for instance, has been studying the brain waves of meditating monks, in close cooperation with the Dalai Lama, without presuming to overturn the Buddhist view that consciousness originates not in the brain itself but in what Buddhists call the "Ground of Being."

Pathological hallucinations, as in schizophrenia, are subjects of study, but a circumspect silence surrounds the cause or correlation of hallucinations in mentally healthy people that occur at times of peril and sorrow. What would prompt normal, functioning brains in roughly half the grieving population to suddenly see or hear things that aren't there? We don't know. Not yet, at least. As the neurobiologist Patricia Boksa notes, "It is unknown at present whether hallucinations are generated by similar mechanisms in patients and in healthy people."

Surprisingly, we also don't know what mechanism creates hallucinations in mentally ill patients. "Neuroimaging data have confirmed the expectation that hallucinations involve altered activity in the neural circuitry known to be involved in normal audition and language," Boksa writes. In other words, unlike imaginings, auditory hallucinations can be shown to *physically involve our hearing.* "However, the major question of how this altered activity arises is still unanswered. . . . In studies with human participants, nueral processes can only be shown to correlate with, not to definitively cause, hallucinations."

The brain remains as mysterious as the deepest fathoms of our oceans and the vastness of space. Britain's National Health Service gave King's College, London, research funds in 2012 to orchestrate collaborative research between experts in psychiatry, neurology, and ophthalmology "to better understand what causes" visual hallucinations. With the current limits of brain science, scientists can map bits and pieces of complex auditory and visual hallucinations and are eager and curious to figure out the whole picture. But in the meantime, describing a sensed presence as a hallucination is presumptuous and unhelpful. As the physicist Harold Puthoff said, it isn't what we know that gets in our way, it's what we believe.

Martha Farah, director of the Center for Neuroscience & Society at the University of Pennsylvania, has said it best: "We should cultivate a certain epistemological modesty," she told colleagues in a 2009 lecture, "and not assume that we can explain everything that matters—or even what it means to matter—in terms of chemistry, biology, and physics. And certainly, we should not infer that whatever cannot be explained in those terms does not matter."

People who experience the Third Man draw strength and guidance from the mystery, and have a sense of being cared for and watched over. This is not dissimilar to people who sense the presence of the dead. Indeed, sometimes the two phenomena overlap, as in William Bird's story from World War I, suggesting that the category distinctions are superficial, even if one experience is scientifically attributed to oxygen deprivation, say, and the other to loneliness and grief. The American adventurer Ann Bancroft, who was completing the first all-female land crossing of the Antarctic in 2000, was lagging behind her travel mate as she tired, when she encountered "an abrupt sense of being in very close company with another person," whom she realized was her deceased grandmother. It wasn't the grandmother she had been close to in life—not the one she might have expected. Nevertheless, she was all at once "infused with a sense of comfort, warmth, and strength," she told the journalist John Geiger. "It startled me because there was a flood of emotion with it, because it was so strong, and it was good medicine, it was what I needed."

The astronaut Jerry Linenger was working on the space station Mir in 1997 when he became aware of the presence of his father, who had died in 1990. Linenger spoke to him, and felt uplifted and moved by the "visit." His father conveyed his pride that his son had realized his dream of traveling into space. Later, Linenger chose to interpret the presence as a projection of his imagination, nothing more. And yet at the time, he derived great consolation from the encounter.

In the grief literature, sensing the presence of someone deceased has been defined as "clearly seeing a figure of a human form, someone who was not physically present at that moment," or "vivid sensations of some presence, as if someone or something touched or pressed on all or some part of the body." Visual perception of the presence seems to be the rarest. Only about 5 percent in one study actually saw the deceased; auditory perception makes up about 15 percent; and the rest are partial impressions, like my sister feeling hands on the back of her head and noticing a distinct presence in the room.

The New Mexican writer Nancy Coggeshall told me about feeling the presence of her deceased partner, the rancher Quentin Hulse, five months after he died in 2002. "I felt pressure on the mattress beside me in bed. The second time the pressure is so great I roll over to see who is there." Hulse, locally celebrated as one of Gila County's last true cowboys, returned to her four years later. "So strong I woke up and asked who's there? I swept my hand over Quentin's side of the bed to see. This was the second visit. So strong was the sense of his being there I felt someone lying down beside me. FELT the impact of weight on the mattress. (Both times.)"

The prevalence of this experience ranges across cultures, but seems to touch, on average, about half the bereaved population. In a 2006 study, it was found that in 86 percent of cases the sensed presence was the first revelation that a death had even occurred. Of those interviewed, 84 percent were in good mental health at the time of the encounter. The time of day, and the level of light, made no difference. Only 5 percent found the encounter to be negative or distressing. For the majority, it was profoundly comforting.

For most of the twentieth century, the model in grief therapy was to encourage people to "let go," to seek closure, to give up their "neurotic" attachments to the dead. In the context of recommendation to "break bonds," sensed presences were first characterized as pathological hallucinations. Sigmund Freud's 1917 essay *Mourning and Melancholia* described healthy recovery from loss as the successful severing of ties. Loved one: exit stage right. Those who sensed presences were, in Freud's view, "clinging to the object through the medium of a hallucinatory wishful psychosis." Not only were the presences and voices and touches not real—but they were *unhealthy*. It was a sign of therapeutic progress when a patient "gained insight" into the fact that they were imagining things. Freud was able to theorize freely about wishful hallucinations because there wasn't a neuroscience, yet, to test his claim that the brain was capable of conjuring such visions at will. Now that we have a better-developed neuroscience, we haven't proceeded to test his claim but, instead, *use his claim* to validate current psychiatric thinking about grief.

Freud promoted his point of view even as young soldiers were experiencing guiding and helpful, even lifesaving, presences on the front. Ironic, that while they were coping and surviving thanks to what they perceived as deceased loved ones and guardian angels, their mental states were being stigmatized for such perceptions by the armchair analysts of Europe.

Here is what medical students typically learn about grief hallucinations in their textbooks:

"The hearing or seeing of a close, recently deceased friend or relative is not a mental disorder," explains one Intro to Psychiatry book. "Usually these hallucinations become less frequent and cease over weeks or months." (Actually, according to bereavement counselors, they can go on for years, or appear *after* many years. In a study of surviving AIDS partners in San Francisco, 22 percent were still sensing or seeing their beloved three and four years later, and the majority of these reported a deepened sense of spirituality as a result.) Continues the textbook: "They are comforting and benign. Perhaps they have a role in helping the individual adjust to the loss. While there is no clear evidence that persons with limited social supports experience more hallucinations during bereavement than those people with extensive social supports, such a finding would not be unexpected."

People held in solitary confinement have been found to hallucinate. Random, often paranoid imagery accompanies agitation, panic attacks, and general mental disintegration after days upon days in total isolation. But this is not what is happening to widows and widowers. They aren't left alone in tiny windowless cells as their psyches disintegrate. It remains totally unclear at any rate how the act of longing could cause a hallucination. Why don't we hallucinate longtime lovers and partners who leave us heartbroken but don't die—people who just go off to forge new romances? Do those kinds of sorrowing visions exist?

In 2008, the psychiatrist Vaughan Bell wrote an article about grief hallucinations for *Scientific American*, pointing out how common they were. He ended by writing: "We often fall back on the cultural catch all of the 'ghost' while the reality is, in many ways, more profound. Our perception is so tuned to [our loved ones'] presence that when they are not there to fill that gap, we unconsciously try to mold the world into what we have lived with for so long and so badly long for."

That's gracefully worded, but we don't know enough yet to assess whether it is true.

Research has shown no consistent connection between levels of social support, education, or even religious belief systems in people who perceive the dead. Here is an account from a lawyer, interviewed by the psychologist Erlendur Haraldsson in 2006, that shows why a longing for the lost one isn't, in itself, the *cause* of what we see. "I was coming home from a dance. I had not tasted a drop of alcohol," the lawyer recalled. "It was about four o'clock in the morning and full light as we were in the middle of summer. I was walking over a bare hill on my way home from town. Then there comes a woman towards me, kind of stooping, with a shawl over her head. I do not pay any attention to her but as she passes me I say 'Good morning' or something like that. She did not say anything. Then I notice that she has changed her course and follows me a bit behind. I got slightly uneasy about this and found it odd. When I stopped, she stopped also. I started saying my prayers in my mind to calm myself. When I came close to home she disappeared. I lived in a house on the compound of a psychiatric hospital where my father worked. I go up to my room. My brother wakes up and says half asleep, 'What is this old woman doing here? Why is this old woman with you?' And I tell him not to speak such nonsense and to continue sleeping, although I knew what he meant."

In the morning, their father told them that one of his patients had died at three that morning. "What I had seen," the lawyer—well familiar with the challenges of eyewitness testimony—told Haraldsson, "fitted her description perfectly."

Whatever drove this man's encounter, it wasn't the poetry of longing, and it wasn't the crisis of isolation.

In 1994, advertising professional Karen Simons lost her father, unexpectedly, to a heart attack. The events surrounding his death show, again, the fluid overlap in experiences that theorists try to keep separate. It took me a couple of years to get her to talk about her experience, after she'd conveyed it to me in passing one day. A tall woman with striking green eyes, she sat in her spacious office with her elbows on her desk, the company key lanyard around her neck, paper and books in piles all around her. Her laptop pinged almost continuously as new emails arrived, but she ignored them in order to focus on the events surrounding the death of her father. A farmer who had just turned eighty, he was still hale, with many plans, although he had had heart problems. In the autumn he'd flown out to the West Coast to visit his grandchildren, and while there, one of the granddaughters had a dream.

"I had a dream where he asked me to take him home, that he couldn't die here in Seattle, that he needed to get back to the farm.

I remember the dream being outside, with miles of hills and forests and roads around, and he was very weak looking. He kept asking me to take him home but I couldn't figure out how to get him there. I asked him how and he said I could carry him, but I wasn't strong enough and I told him so. He told me that he really needed to get home and knew I could carry him there if I really tried. So I boosted him and started to walk."

It's a lovely image, and it reminds me of something a palliative care counselor once said about what we fear most: "Who's going to carry us when we die?" Who is going to remember us, to know we're still there, to keep us from feeling alone?

"Dad arrived home [back East] on a Saturday," says Karen. "On Sunday night, he insisted on cooking dinner for us," for her, her husband, and her sons, "and the next day he was going to a farmers' convention. The last sight of him I had was early in the morning; he waved at me and woofed our newspaper close to our door with an overhead throw." She mimics the gesture. "That night, on the way to the convention banquet, he had a massive heart attack and died.

"After Dad died, I began driving his big, old Ford Taurus. It was comforting, in a way, the way you hang on to people's shirts. But that's all it was. Until about six weeks after he died." She tilts her head and gazes past me, furrowing her brows as she tries to recall the exact timing: "It was a very cold night in January. I'm driving on the highway, and into the passenger seat comes Dad. I could feel him settle in. He had a very distinctive lean to the left, because of the way his back was. Also, you know how you know the sound of people's breathing? How you can tell whether it's your son or your daughter in the room? There was Dad. He rode with me from about Kennedy Road to Pickering. [Ten miles.] It was incredibly real, and it was completely transforming. I was almost giddy. I was hoping he would stay."

She never sensed him again, and decided that he had, in effect, been taking his leave. She remembered speaking with a Buddhist friend who explained their view that the soul lingered for forty-nine days, or roughly six weeks, before departing. That, however, wouldn't account for her aunt, whose son died thirty-five years ago. "He was a heavy-equipment operator who went through river ice into a deep and fast current. They never found him, or his machine. On a very regular basis, Allan comes and sits on the end of her bed, and they have a conversation. *And don't tell me I'm crazy!'* my aunt always says." Simons folds her hands beneath her chin, leans on her desk, gazes at me, and laughs. This is just the way life is, she is saying. Our families are full of ghosts. (Interestingly, Simons comes from Norwegian and Scottish ancestors. A study of sensed presence experiences in Norwegian widows and widowers found that an unusually high number—75 percent—reported having one.)

Surveys of people's reactions to grief counseling show, unsurprisingly, that they often feel "unaccepted, abnormal, not understood," when they relate these encounters to healthcare professionals who aren't comfortable with the spiritual implications. The Australian palliative care physician Michael Barbato, moved and astonished by the uncanny experiences he repeatedly encountered in his patients, wrote that, "Only after being informed about the commonness and normality of post-bereavement hallucinations did most other widows and widowers speak freely, expressing relief from thoughts that they might be considered insane."

Why should you be considered insane for encountering something so lovely? Here is an account from the Scandinavian writer Johan Kuld, about an experience he had shortly after the death of his wife. It is a tale that one can surely only envy:

She came to him, pushing open the door of the room, smiling. "I stared at her as if I was hypnotized and could not say anything." She felt warm, normal. She explained that she was there to say good bye, and bid him lie down on the bend with her for a time. "I did as she asked," he wrote. "She stroked my cheek and whispered beautiful things to me. And I was filled with a feeling of joy. Then I was filled with a calm and I was surrounded with a sense of peace that cannot be described with words." He sank into sleep as she caressed him, and when he awoke, she was gone. "This incident is still so vivid in my mind. It does not get old or fade with the years that have passed since then. I have often asked myself: 'Could I have dreamt it all?' But my answer is a definite: "No." This was reality as far as there is a reality, unusual and unforgettable at the same time.

That a large percentage of people sense, see, or hear from the deceased would not have surprised most humans throughout history, who lived out their lives with the bedrock assumption that the dead continued watching, consoling, guiding, and even meddling in their affairs. From Rome through to nineteenth-century Iceland, the living shared their pastures and roads with those who'd predeceased them. They still do in parts of Indonesia, Thailand, and Japan, among other places. Everybody could potentially see the dead, and a huge amount of effort was invested by communities in preventing the spirits from wandering or getting lost on their way to their appropriate destination in the afterlife. Certain spirits were at risk of haunting the living because of the tragic or violent nature of their demise. To keep them at bay, in some societies bodies were decapitated before burial or tied firmly to the ground. Suicides and criminals were secured beneath boulders, or burned. Some were thrown into bogs and rivers. It wasn't only Bram Stoker's Dracula who needed to be mutilated before his *un*death became a more reassuring true death.

As the French medievalist Claude Lecouteux notes, "The world was haunted, by the dead transformed into spirits passed on to another state, living another life in permanent conjunction with 'contemporary' humans and always capable of either giving them information or ceding to their imperious requests. The people of the Middle Ages had no fear of death: they dreaded the dead—some of the dead, in any case."

The English historian William of Newburgh wrote in the 1190s, "one would not easily believe that corpses come out of their graves and wander around, animated by some evil spirit, to terrorize or harm the living, *unless there were many cases in our times* [italics mine], supported by ample testimony." Given what we know now about the prevalence of deathbed visions, grief hallucinations and Third Man appearances, you can't accuse our ancestors of being "unenlightened." They saw what they saw and it was no different from what some of us see.

In Europe, the church was uncomfortable with this lived reality of sensed presences because it didn't dovetail properly with Christian doctrine. "Augustine posed the problem of perception," writes Lecouteux, referring to the influential fifth-century theologian. "Are these apparitions the creations of slumbering, somnolent, or feverish men?" Upon reflection, Augustine decided that the apparitions were illusions or messages projected into the psyche by God: "It communicates to the dreamer that these dead bodies require burial," for instance, "and all without the knowledge of the former owners of these bodies." God acting as a ventriloquist? Fifteen hundred years later, those scholars interested in the subject continue to speculate that a combination of things may be going on here: spirit, imagination, dream, source signal. The mystery continues.

Chaz Ebert, the Chicago lawyer and widow of the film critic Roger Ebert, awoke the night that her first husband's father died to an almost cartoonish encounter with the grim reaper, as if her response to the faint signal she was picking up was to engage in a childlike dream. "This sounds really crazy, I know," she told the historian Studs Terkel. "I *never* believed in the grim reaper, I thought it was just some mythological thing." And yet, one night, she awoke in the darkness, opened her bedroom door, and encountered him. "It had the monk's robe and the whole thing," she said. No face, "lit orbs where the eyes should be." Somehow, telepathically, the creature conveyed that he wasn't there for her; instead he bade her look back toward the bed, where she saw a coffin, in which lay her father-in-law. "This lasted for a few seconds, then it all disappeared." She roused her husband to tell him what was happening. He said, 'I think you were dreaming.' I said, 'No, I *wasn't* dreaming, I'm awake!' So we go to sleep, and the next morning we call and they say his father died. His father died at the same time that the grim reaper came to my house."

Did the grim reaper literally come to her house? No. But the news of her father-in-law's death did.

Of such phenomena the scientist Dean Radin writes: "That vision is a construction from your memory and imagination, similar to a waking dream, except that the stimulus for the image is occurring somewhere, or some-when, else." Perhaps, speculates the psychologist Erlendur Haraldsson, "The deceased moulds the perception in the mind of the living person." Again: like the ringing alarm clock piercing your sleep and becoming a telephone or a church bell in your dream, there is an instantaneous collaboration between world and mind.

Pope Gregory the Great (540–604) pursued the quandary of apparitions down a path that would ultimately contribute a little to the establishment of Purgatory as a distinct location between heaven and hell, where restless souls awaited assignation. "Henceforth these individuals were regarded as imprisoned souls, or the damned," writes Lecouteaux. "It was believed that the elect show themselves at times, but they are easily identified because they are radiant with happiness and beauty, and the clerics were incapable of confusing them with the sinners suffering the punishments of purgatory."

Eventually, what seems to have happened is that the Christian church reshaped revenants from guiding ancestors and the disoriented or vengeful dead to spiritually suspended beings who depended on the living to pray for their passage to heaven. This reshaping enabled the church to take control of pagan ancestor worship and bend its purposes to Christianity. Hence the monks of Cluny established November 2 as All Souls' Day to incorporate the pagan feast of the dead. The irony of what the church accomplished, though, is that it managed to take common and powerful experiences, each of which has its own context—the Third Man in times of peril, the deathbed apparition, and the beloved deceased in our ongoing lives—and turn them into church-related set pieces that nobody believed. In other words, when belief in church doctrine in general declined in the West, so did belief in experiences that were never *inherently* connected to a particular religion in the first place. This distortion never happened in Asia, where the dead are still freely interacted with and honored.

There is a kind of sensed presence that is not helpful or beloved but, instead, is completely unnerving. It's worth digressing into this territory because it may explain why people throughout history have gone to so much trouble to quiet the troubled dead. This is the terrifying specter that assaults some of us during episodes of what has come to be called, in medical language, sleep paralysis.

David J. Hufford, professor at Penn State College of Medicine, is the world's leading expert on this phenomenon. Hufford discovered it through his own out-of-the-blue experience in the 1960s. "I was a sophomore in college. I'd gone to bed early after my last final exam. I was very tired. Being sleep deprived makes this more likely. I woke up hearing the door of my apartment open, and I thought it's probably somebody coming to see if I want to go for dinner. I heard footsteps coming across the room. And then I found I couldn't move, and that was very frightening. Then I felt the bed press down as if somebody was climbing up on it; then I felt what felt like someone kneeling on my chest, and then I felt hands on my throat. I thought I was being killed. It was horrible. I had a feeling of revulsion and terror about this thing on my chest, not just because it was trying to kill me, but also because it seemed evil. I struggled to move, and when I did move it was gone. I leaped out of bed and nothing was there. That is a very typical sleep paralysis experience."

Approximately 30 percent of the population has had a simple episode of waking up feeling briefly paralyzed, but encounters with the malignant sensed presence happen to 3 to 6 percent, according to the psychologist J. Allan Cheyne, of the University of Waterloo. It was initially assumed that episodes of sleep paralysis were colored by cultural expectation, that if you roused in this peculiar, halfway state between sleep and wakefulness and expected to be visited by a werewolf, a witch, or whatever your culture believed in, then you would be. However, people like Hufford, who had no prior knowledge of the phenomenon, can experience it out of the blue. Another such person is the novelist Barbara Gowdy.

"It began in my early twenties," she told me. Gowdy is a beautiful woman: physically graceful, with rich brown eyes and a fierce intelligence. She questions everything, holds every object up to the light, and examines every facet. This experience was no exception. "It only happened when I took naps, and if I was lying on my back. The first time it happened, the room got very cold, and very dark. I could hear my husband in the kitchen [but] I was suddenly aware of a hideous, unspeakable evil in the room. And then I was aware that it was on my chest. It was so real, and concrete. And I couldn't move or call out. Suddenly, as if waking up, it would be gone."

Twelve or fifteen years later, it started again, also during a nap. "I had my eyes open. I hadn't gone to sleep yet, and the temperature changed again, and the evil presence was back. It's the complete opposite of bliss, or of the feeling you have with a newborn baby. This evil was deeply old. It wasn't the evil of a brand-new psychopath. It was evil that had been in the world since time immemorial—a wise, experienced evil. It gave me access to an idea of evil that I hadn't had before. It was dirty—and old—and sexual. [Experiencing] it was a fate worse than death."

Gowdy went to research what was happening to her in the public library, although there was scarcely any information available. Until the late 1970s, the sensed presence in sleep paralysis was considered to be a rare side effect of narcolepsy. "I saw artists trying to capture the evilness, but no one could quite capture it. Until you've had the experience, you can't know it."

Desperate to avoid her tormentor, Gowdy stopped having naps. For years she managed to evade the ineffable terror. "But then it started happening again. This time, it whispered, low and growling, in my right ear: 'I am the wolf.'"

Fed up with being terrorized, Gowdy abruptly and angrily surrendered. "Well," she replied, "do your worst."

At this, "the hallucination popped and disappeared and never came back. That suggests to me that I actually took on my own subconscious and, by fearlessness, I beat it." She also wonders whether it wasn't a deeper, collective unconscious she was tapping, or as she put it, "the memory of pain, the memory of evil."

Gowdy's reflection reminded me of a time about a week after I interviewed Audrey Scott, the self-published author who died in her home. I woke up at 3:30 a.m. beset by troubled thoughts, haunted by a book I'd just been reading called *The Tiger*, by the journalist John Vaillant. A man in some hellishly barren and isolated town in far eastern Russia had hanged himself with his belt after his only son was eaten—nothing left but boots—by a solitary Amur tiger. I awoke feeling intense empathy for this father, fully inhabiting the horror of living with his knowledge, that his child had become meat and there was nothing else—no job, no love, no prospect—to ameliorate that soul-destroying truth. I tossed and turned in the airless bedroom, drifting along the shoreline of sleep. And then all of a sudden I was in . . . a lucid dream? A night terror? I felt as if I'd fallen into a force field, like a wind or electricity field, or at any rate a realm, a *living realm* of dread.

Maybe I simply felt intense dread in response to the unfamiliar sensation, but I'm certain that within it, I was powerless and trying to hold on to myself, talking to myself in my head, saying something along the lines of "Stay strong, you can survive this." Although I can't remember now exactly what I was saying, I was trying to separate myself, the backbone or essence of myself, from this force. It wasn't dissimilar to the effect that Tolkien described when the hobbits foolishly gazed into Saruman's crystal, and beheld the eye of Mordor.

There were no physical effects, the way there is with a panic attack. My heart wasn't racing. I just wanted to be free of the experience. At last I was able to turn on the light and listen to a talking book for an hour until I felt it was safe to try falling asleep again.

"We have been especially struck with the frequency that the specific term 'evil' is applied to this presence," writes Allan Cheyne about such episodes, "even by people to whom this term does not readily spring to mind." As one of his study subjects said: "I literally fear for my soul."

How does this sensed evil feel different from the perception of threat, as in a feared attack by a wild animal, or a hurtling car, or an impending tornado? Consciousness discerns a difference between imaginary threats of one kind—impersonal, animal, weather-driven, or even a palpably insecure and jittery assailant—and a threat that is *powerfully evil*. There has to be a quality to that perception that extends beyond what is generally perceived by the threat-scanning function of the amygdala in the brain. But no one seems to know what it is.

These terrifying presences materialize only when people are dozing off, it is argued, because they inhabit the realm of hypnagogia—*hypnos* is sleep and *agogos* is "leading in"—that strange layer of consciousness between sleeping and waking that often features whacky thoughts and images. Research at Fukushima University in Japan has established that people experiencing sleep paralysis show an EEG signature that mixes wakefulness with REM sleep. So, we are speaking of dream physiology, but not of typical (which is to say, wildly idiosyncratic) dream content.

"Hypnogogic experiences are a bit of a mystery in the sense that scientists don't know exactly how to classify them," writes Jeff Warren in *The Head Trip: Adventures on the Wheel of Consciousness*. "Are they dreams, or thoughts, or something else entirely? Where exactly do they come from? Is there any logic to their appearance?" It's an important question, particularly in light of the research suggesting we are more susceptible to telepathic impressions and other uncanny perceptions when we have dimmed down the other noise in our conscious awareness.

Most dreams and hypnagogic images are unique to the dreamer, but the more consistent experience of hearing a door open, hearing footsteps, and sensing or seeing a deeply evil creature that crushes or even sexually assaults you makes sleep paralysis puzzling. Why not react to the sound of crickets outside your window or the distant sirens on a city street? Why not incorporate the sensory input that is a houseguest flushing the toilet or a neighbor having a party? Why does the door open? Why do the footsteps come? There is a medieval account from Sweden where the presence is said to sound like a cloth sack dragging across the floor. The Navaho have heard moccasins shuffling; patients have heard stockinged feet on hospital carpets.

Then there is the dread beyond words. "The greatest primal terrors that I have ever witnessed: character-forming stuff," says one person. "These attacks leave me shuddering and crying," says another. "Sometimes I'm so scared I get sick to my stomach."

Very commonly, there are auditory hallucinations. For one man, these have included a growl that was hellish beyond reckoning: "To describe the demonic growl as the epitome of evil would be an understatement. It was pure, unadulterated malevolence. It struck fear into the very core of my being, almost shattering my soul. I hope to God that I never hear it, or anything like it, ever again."

"The complex pattern of sleep paralysis, including the evil presence, the shuffling footsteps, and a host of other details, does not arise from culture," David Hufford has argued. "Rather, in cultures around the world, traditions of spiritual assault arise from this experience." Thus we have the succubus and incubus of medieval Europe, or the "sitting ghost" of China. The experience is the origin of the word "haggard" which began as "hag-rid," or being ridden by the Hag. Similarly, "nightmare" comes from the Anglo-Saxon word "mare," or "crusher." Nightmare: someone who crushes you in the night. There are occasions when the experience is "intersubjective," meaning that it happens to a number of people in, say, a village during the course of one week, or to two people in the same room. I spoke to a friend who recalls waking up with the sensation of being paralyzed while in a country house in England in the early 1970s. She had an apprehension of pure dread. She heard a baby crying and crying, a desperate, needful sound in the hollow dark. Her lover lay beside her and had the exact same experience at the same time. He was also paralyzed; he also heard the wailing child.

"The neurophysiology of the paralysis is very well understood, we know the biochemistry and we know the pathways," says Hufford. "But we don't know any pathway that produces *all* of this. We can say that in the sleep paralysis state, with that physiological trigger, you're in an altered state of consciousness. So far so good, but that's as far as the conventional explanation takes you . . . I don't know any explanations that fit the data, other than the traditional ones, which is that it's a spiritual experience."

Hufford notes that when a group of psychiatrists was presented with a sleep paralysis case study and asked to diagnose the patient, over half labeled the anonymous subject "psychotic." As with the other sensed presence encounters by the grieving, lost, or stranded, this is the label they receive—and fear.

When Hufford set about researching the subject, people often admitted that he was the first person they'd told. "One of the most fascinating social issues here is that all of these kinds of experiences were well known in Western tradition up to three or four hundred years ago. It's not simply that Western culture never had a clue about these things. We erased knowledge of these experiences from the cultural repertoire while the experiences were continuing to happen. That's a level of social control that's very impressive." Perceiving a spiritual being, whether loving or cruel, has become, Hufford proposes, "an illegal experience."

This prejudice is beginning to change, particularly in the area of grief therapy, as counselors start to take note of other cultural approaches. There continue to be societies throughout the world that sense or see the spirits of the dead and accept them for what they appear to be. One influential study of Japanese widows found that their continuing bond with the presence of their deceased spouses in the context of Japanese ancestor worship—setting up altars in the home, leaving food, treats, incense—made them much more psychologically resilient than their American counterparts. The people they love brush their shoulders, their cheeks, stay among them, and are nourished in turn. In the country of my birth, Mexico, Halloween is still trumped by Día de los Muertos, or the Day of the Dead, and families picnic in the graveyard. They lay out favorite food and drink, and festoon the tombs with marigolds while shopkeepers sell prancing skeletons. Death is in life; it is part of life.

I find myself envious.

London neuropsychiatrist Peter Fenwick has commented on this: "Often—and for the people concerned this is the most significant aspect of the experience—its emotional impact is so great that it remains a lasting source of comfort to the recipient and often has the power to alter their own perception of what death means. For them, whether what happens is dismissed by others as 'simply coincidence' is irrelevant: the fact that it happened is enough."

Sometimes it's enough, but often it is not. We are held back from embracing the comfort and reassurance of spirits by a society that belittles the experience, that says "show me the money, which is to say: the proof." There is no proof. Neuroscience simply cannot tell us, at this juncture, why people have "sense of presence" experiences. What used to be known as guardian angels, succubi, and ancestors are now simply called hallucinations. The shift in semantics is fine for some people, like the astronaut Jerry Linenger, who figured his visiting father was an illusion but felt consoled nonetheless. For others, however, it creates an enormous loss of meaning. "There is a danger that in objectifying or analyzing an experience we may lose sight of its significance for the bereaved or the dying," the palliative care doctor Michael Barbato has ventured. The best thing that ever happened to you becomes the thing that others want to deem unreal. *I don't mean to be unkind, but your sister was clearly imagining things.*

"Where is the wisdom we have lost in knowledge?" T. S. Eliot asked in his poetry almost a hundred years ago now. "Where is the knowledge we have lost in information?"

We need to accept what the bereaved—and the terrorized and the stranded—do or don't see. They are entitled to forge their own meanings.

CHAPTER 5

Be Still: How the Dying Attain Peace

Bleak midwinter in northern Minnesota: a frozen forest, hushed and empty, with scarcely a road cutting through the miles of spruce and pine trees. People, for the most part, traverse it high above in jet-liners, or—if they are making their way from one regional spot to another—shakily in twin-propeller planes. In January 1979, a young doctor named Yvonne Kason was traveling in this regional fashion, evacuating a critically ill Ojibwa woman with measles encephalitis from a tiny, fly-in reserve to the nearest hospital, a few hundred miles away. Traveling with Kason was a nurse, Sally Irwin, and the pilot, Gerald Kruschenske, who was confident he could keep his twin-engine Piper Aztec on course even as the wind began to rise and snow swirled in from the west. The passengers hunkered down, tense and cold, as the weather intensified into a blizzard.

Over the course of some horrifying moments, ice began to film the plane, and the right propeller blade faltered. Then the left pro-peller sputtered and stopped. The Piper began to lose altitude, and when it became apparent to the pilot that he couldn't clear a loom-ing hilltop, he had no choice but to try to glide into a crash landing.

"I felt intense panic, intense fear," Kason told me thirty years later. "I thought, 'Oh my God, I'm going to die.' But it came out fast; it was like a death prayer, *ohmyGodI'mgonnadie*—this call from my soul for help." Immediately, to her immense and lasting surprise, she felt a wave of tranquillity descending, washing out the terror, which receded to leave only peace. "My mind became calm, and I was no longer afraid, and I was alert, and conscious like I am right now." Perhaps a second or two had elapsed, no more. Then, again to her astonishment, she heard a voice. "Be still," it said, "and know that I am God." Born into a Christian family, Kason hadn't been to church since she'd participated in a United Church choir in adolescence, but she was clearly remembering, or somehow hallucinating, a verse from the Old Testament's Psalm 46. *"God is our refuge and our strength, a very present help in trouble,"* the psalm begins. Relax, fear not, be still. "I am with you, now and always."

"What do you mean by 'the voice'?" I asked her, pen in hand, as we sat kitty-corner at her dining room table drinking coffee. I had sought her out after hearing about her experience from a psychologist at the University of Toronto. I wanted to know why my sister had seemed so serene and unafraid at the end of her life, as if she were surrounded by or encountering an invisible source of joy. In Karlis Osis and Erlendur Haraldsson's deathbed research, they found 753 cases in their American and Indian surveys in which the dying experienced an "elevation in mood" right before they passed away. Eighty-seven percent of those who saw visions with a "take away" purpose—someone beckoning, for example—died within the hour. While most of the patients who had the "take away" visions reacted by becoming elated or serene, only a fraction—7 percent—responded to other kinds of hallucinations (elephants, swirling lights) with a change in mood. Why this difference?

It didn't seem to correspond to a release of endorphins. "Some patients in our sample received injections of morphine and other substances closely resembling endorphins," the psychologists wrote, "yet afterlife-related experiences were not increased in that group. If these medically administered substances did not induce the phenomena, why should they be the cause when internally secreted by the body?"

There had to be a way to play with this question, I thought. Could someone who hadn't died, yet who'd come to that very edge, shed light on why some of the dying might become radiantly peaceful?

For Kason, it was the experience of the unexpected voice that began, at first, to calm her. "How was it not your own voice or thought?" I pressed. She considered the question, pushing a pale strand of hair behind her ear. Elegant, intelligent, warm, and engaged, Kason is in her midfifties now and understood why I was pushing her for more detail. In medical circles, you don't get to just claim that you held yourself together during a plane crash because a voice commanded you to calm down. "Okay. It's like you're walking down a hall, and someone comes on a loudspeaker and tells you to stop. You *know* it's not you." She studied me, to see if that analogy worked. "I mean, it was that clear. I was being instructed."

By whom? A man, a woman?

"If I were to put a gender on it, I would say it was more masculine than feminine. It was a low voice, a deep tone. It was *profoundly* comforting." She spoke confidingly, as if all these years later she still couldn't believe this crazy thing had happened.

Listening to her, I mused: here we have the Third Man in an emergency, only instead of providing guidance, the presence or voice offers strength. Calm. In a rushing matter of seconds, wind howling and plane tumbling, Kason felt herself to be in an altered state of consciousness. "I was not thinking or judging or analyzing, the way one normally does." Nor was she numb, or in denial; it was "the sort of peace one gets in meditation," she explained. "You're in what they call a 'higher mind' state, which is intuitive and receptive. I was watching the plane crash, and the nurse and I were bracing the patient, and I felt profoundly peaceful and calm. I knew there was nothing to be afraid of. I just knew it in my soul. *Even if we were going to die.* And I started comforting the patient, 'cause she had woken up and her eyes were looking at me, as the doctor, and I was able to now speak to her and transmit the sense of comfort. I just kept saying, 'It will be okay, it's going to be okay.'"

That is lightning-quick certainty for a twenty-six-year-old woman who hadn't pondered the afterlife much. "I knew with absolute certainty something I had never known before: that there is absolutely nothing to fear in death." I ran my hand absently along the back of her cat and remembered what Katharine had whispered to our sister Anne in the hospice, how she was no longer afraid. In a recurring dream Katharine had for years, she found herself in a slowly crashing plane, skimming the tops of trees. Always, it was a nightmare. Always, she was afraid. In the hospice, I hadn't been able to understand why she no longer was, and she was too short of speech to elaborate. Was it denial? Evasion? Morphine? An endorphin rush? Or had a mysterious, loving voice commanded her to "be still"?

Kason resumed her recollection, of descending to a partially frozen bay on the Ontario side of the border in an area called the Lake of the Woods. "The pilot, really heroically, tried to do a guided crash landing on the ice," she said. "He almost did it, but the ice was so thin that as soon as the plane stopped, it broke through and sank. We had to quickly try and get out, and I was trying to get the patient out, but I couldn't [undo the straps]." The cargo door, through which her stretcher had entered the plane, was soon submerged. Unable to extricate her, the surviving three fell quickly into the water themselves, weighted down by winter gear, and gazed helplessly as the stretcher-bound patient was lost with the Piper.

The Third Man—not a presence, mind you, just a voice—took charge of Kason once more. "This voice started repeating, 'Swim to shore.' I actually argued with it, because I had taken lifeguarding, and they always tell you, in a boating accident, don't try to head for the shore. It will be farther away than it looks. I tried to ignore the voice, and get on the ice. But every time we put any weight on it, the ice would break off. And when you're cold and wet you get tired *really fast*. My parka and boots were like lead, dragging me into the lake. So I surrendered to the voice, because what I was doing wasn't working, and I started swimming to shore. It was really, really difficult. I didn't think I was going to make it. I kept going under, and the water filled my lungs. Somewhere in that life-and-death struggle is when my consciousness suddenly whooshed"—she gestured with her hands, sweeping them up either side of her head—"and it was like I was no longer looking out of my eyes. I was twenty or thirty feet up, and I could *see* myself swimming." She gazed down at her dining room table, as if still amazed. "This, to me, was very bizarre, because I'd heard descriptions of people going out of body when they were lying down, but not when they were swimming!" She looked up at me and laughed. "I puzzled over it for years."

To this point in her story, what Dr. Kason experienced remains within the realm of comfort for psychologists—they believe they can explain or, at least, theorize about it. Research shows that when people are confronted with their own demise—in plane crashes, car accidents, falls from cliffs, even jumps from bridges (as later reflected upon by the few who survive), they enter a state of dissociation. It is a defensive stance, essentially, triggered by *anticipating* trauma and psychologically fleeing from it, making it seem as if it is happening to somebody else. Different researchers put different spins on this theory. But all agree that certain elements are common.

In the Atlanta-based cardiologist Michael Sabom's 2004 study of accident survivors, forty-three of fifty-two people experienced disso-ciation during the event, suggesting that it is a typical—rather than unusual—reaction to the prospect of death. A flight attendant who survived a United Airlines crash in 1989, told Sabom: "I felt, this is . . . this is it, this is how I'm going to go, this is how I'm going to die, and it was the most incredibly peaceful moment I've ever known, that I was in no pain, I had no fear anymore, it was total peace." This oc-curred even though the flight attendant never physically came close to death, or even serious injury. Sabom decided to dub this extraor-dinary psychological state the "acute dying experience," which is not to be confused with actually, acutely dying (or, as my cousin briefly thought I said when I was talking about my research, a *cute* dying experience. No, not accidentally smothered by puppies, but equally pleasant perhaps).

Sabom argues that his concept of acute dying experience "mir-rors the three-stage response of perception, defense, and recupera-tion observed during predator-prey interaction in animals. It appears to be an adaptive response promoting survival in the acute situa-tion." He points to research on mice. They appear to have a two-tier fear response depending on how extreme a threat is. A sudden bang-ing sound in the kitchen where they've been scarfing down crumbs will put them in flight mode. But mortal threats, such as swoop-ing hawks, "elicit a higher, second-tier response in mice associated with . . . an instant release of endorphins."

If the predator attacks, the mouse needs to concentrate on sur-vival strategies and not be distracted by anxiety and pain. Hence the endorphin flood that would make them pain free and calm. If the mouse manages to scoot away, it can enter the "recuperation" stage, where pain floods in and commands attention to whatever wounds it received. Following this animal model, Sabom and other researchers argue that the perceived severity of the threat determines whether we shriek, freeze, cower, or . . . go into Zen mode. The horror movies have it all wrong. Anthony Perkins attacking Janet Leigh in the shower would not necessarily have made her scream. Instead she would have stood frozen and silent.

In 1893, the Swiss geologist Albert Heim published his exploration of the emotional state of mountain climbers who had fallen in the Alps. Heim, himself, had taken such a plunge, and had been astonished by his radical transformation from terror to "a divine calm (which) swept through my soul." The other climbers he interviewed had similar experiences of transcendent peacefulness. As death appeared imminent, his collection of tumbling men "often heard beautiful music and fell in a superbly blue heaven containing roseate cloudlets."

Some people, like Kason, have the bizarre perception of leaving their bodies. A U.S. marine in a 15,000-foot free fall into the Pacific Ocean due to a failed parachute recalled this startling shift in his perspective: "I was falling and working with the chute and just like that" (snap of the finger) "I was 15 or 20 feet away watching me struggling." He was fascinated by the brilliant orange of his flight suit, the look of his helmet, the unusual boots he'd forgotten he was wearing that day. "I have never," he told Sabom, "experienced (anything like) it before or since."

A second tale from the Alps was written by the amateur Swiss climber J. L. Bertrand, who got caught on a ledge so narrow that if he moved he would fall to his death, but if he didn't move he would freeze. He chose the latter course, and as he got colder, at some point he floated above himself. "'Well,' thought I, 'at last I am what they call a dead man, and here I am, a ball of air in the air, a captive balloon still attached to earth by a kind of elastic string, and going up and always up.'" From his newly expanded vantage point, Bertrand spied his climbing guide, farther down the mountain, drinking Bertrand's Madeira wine and eating his chicken lunch, about which he teased the guide later. He also saw his wife, hundreds of miles away, traveling to the village of Lucerne.

"My only regret," he wrote, "was that I could not cut the string. In vain I travelled through so beautiful worlds that earth became insignificant." When he "returned" to his body after being revived by the lunch-thieving guide, who had found him, "my grief was measureless. . . . I felt disdain for the guide who, expecting a good reward, tried to make me understand that he had done wonders. . . . I never felt a more violent irritation."

The psychoanalyst Oskar Pfister, a contemporary of Freud's, read some of these mountaineering accounts and called the experience of Heim's climbers "shock thoughts." This idea would lay the groundwork for what later became known as "peritraumatic dissociation." The psychiatrist Russell Noyes of the University of Iowa researched such experiences in the 1970s and concluded that "depersonalization as a defense against the threat of extreme danger or its associated anxiety" was the best available explanation. But in Noyes's research, the most common features of "peritraumatic dissociation" are distorted time perception, a sense of surreality, a sense of being on automatic pilot, and confusion or disorientation. What this describes actually happened to me, once, when I was a novice driver who accidentally and idiotically turned left into a phalanx of fast-moving taxis on Broadway at Columbus Circle in New York City. I promptly got hit broadside, and my sense of time slowed; I became preternaturally calm. When my crushed Toyota finally stopped spinning, I primly shifted the gear stick into park. About half an hour later, after the police and paramedics had come and gone, taking my passenger and the cabdriver who'd hit me with them to the hospital, I began sobbing and for several hours and could not get a grip on myself. This was the delayed emotional and physiological response that is said to occur with peritraumatic dissociation.

In Sabom's accident study, however, there was a difference between how people reacted when they were suddenly in danger versus how they reacted when they *recognized* the almost-certain prospect of dying. The first group "frequently described an acceleration and sharpening of physical, mental, and visual perceptions consistent with the psychological state of hyper arousal." The accelerated perception, for instance, is what causes time distortion. For Sabom, this feature distinguished his concept of ADE from Noyes's idea of peritraumatic dissociation. He argues that the reactions worked along a continuum, like the first-tier and second-tier responses of mice. Some people merely felt depersonalization, as I did, but "when fear and horror were present, these were always the initial emotions, were short-lived, and were followed by a feeling of peace of *equal or greater intensity.*"

"Considered together," Sabom writes, "these findings present an interesting paradox: the more terrifying and traumatic an accident may appear, the more peaceful and painless the accident may be experienced." (Witnesses, take heart: it is apparently more upsetting for the bystander than for the person undergoing the tragedy, a fact often lost in depictions presented by novelists and moviemakers.)

But if we, like the mice, are capable of tailoring our response to the severity of the danger we're in, does that account for the entirety of all these experiences? Perhaps we are both mice and men. We are capable of having both a physiological and a spiritually transcendent reaction to danger at the same time. There is no other way, at least for Yvonne Kason, to account for what happened to her next.

For about an hour after the plane went down in the Lake of the Woods, Kason found herself suspended between the earthbound and the ethereal, with part of her awareness engaged with the effort to keep swimming, in the iron grip of ice water, while the blizzard blurred her sight, and part of her awareness shifted to what seemed to her an infinitely beguiling light. It was as if the day were morphing in some impossible way, all at once storm and radiant sun. She found herself encompassed and somehow absorbed in the light. She could feel the pain being inflicted by the cold, which an endorphin rush would have blocked. The light was well beyond anything she had ever encountered in her life.

"The experience—it was formless," she said in her airy dining room in a house she shares with her teen son. "It was like dissolving into the light." She considered that description for a moment. "Yes, that's it. It was like dissolving. I was like a drop of water, which had now merged into the sea of light. I still existed, it was still me, but I was in this incredible ocean of light and love." It wasn't just a visual perception of light. It was an experience of full immersion in a *sentient, emotional* light. "The strongest aspect for me was the love. Perfect love. It's impossible to describe."

Ineffable love. Something *you* love, or that loves you? People who take heroin or opium will describe an impersonal sense of bliss, like they don't have a worry in the world. They don't remark on a relational dynamic—an *active love*. But people who witness this light are veritably shattered by the emotional quality it has. Recalling it, Kason closed her eyes and smiled, and even, unconsciously, hugged herself in her white sweater as she spoke: "It was a maternal love.

Like I was a newborn baby on my mother's shoulder, utterly safe." Then she added more shading, for it was not only an infant bliss but also a revelatory feeling: "It was like I'd been lost for centuries and I'd found my way home." Was that what my sister was immersed in as she lay so peacefully on her bed in the West Island Palliative Care Residence? Is *that* why she wasn't afraid?

Light is the core, the essence of experiences that are mystical in nature. "In accounts of the light, contemporary testimony bears a striking resemblance to medieval narratives," writes Smith College religious studies scholar Carol Zaleski. "Both medieval and modern descriptions of otherworld light blend visual qualities such as splendor, clarity, and transparency with sensory/emotional effects such as warmth and energy." There is a synesthetic aspect to how people encounter this light. Its strange beauty is "one of the few truly 'core' experiences," notes Zaleski, "that cut across cultural and historical boundaries."

It is also a central experience in every religious tradition:

Spoke Zoroaster: "Righteous souls will enter heaven, and there they will have a vision of God, who is depicted as pure light."

"The Lord," says Isaiah in the Torah, "will be your everlasting light."

In Hinduism: "Place me in that deathless, undecaying world / Wherein the light of heaven is set."

Jesus: "I am the light of the world."

In Buddhism: Clear Light, and Infinite Light.

Said Abdu'l-Baha, son of the Baha'i prophet Bah'u'llah, in nineteenth-century Persia: "That divine world is manifestly a world of lights."

The sixth-century pope Gregory was obsessed with collecting firsthand accounts of spiritual encounters and reported what he'd heard from witnesses of the light in a series of manuscripts he called his *Dialogues*. "Anyone who has seen a little of the light of the creator finds all of creation small, because the innermost hidden place of the mind is opened up by that light, and it is so much expanded in God that it stands above the world. In fact, the soul that sees this is even raised above itself. Rapt above itself in the light of God."

About a thousand years later, the accomplished and respected sixteenth-century Spanish abbess and mystic Teresa of Ávila, wrote, "The splendor is not one that dazzles; it has a soft whiteness, is infused, gives the most intense delight to the sight, and doesn't tire it; neither does the brilliance. It is a light so different from earthly light that the sun's brightness that we see appears very tarnished in comparison with that brightness and light represented to the sight, and so different that afterward you wouldn't want to open your eyes. It's like the difference between a sparkling clear water that flows over crystal and on which the sun is reflecting and a very cloudy, muddy water flowing along the ground."

Six hundred years after that, hospital patient Monique Hennequin described the light to her cardiologist, Dr. Pim van Lommel: "This luminescence consisted of a kind of infinite river of brilliance, like the brilliance of a setting sun reflected in rippling water with little pinpricks of light like small stars. The brilliance was made up of beautiful little globules of light, extremely bright and quite unlike anything on earth." Another patient of Lommel's exclaimed, "Too much! It's simply too much for human words. The other dimension, I call it now, where there's no distinction between good and evil, and time and place don't exist. And an immense, intense pure love compared to which love in our human dimensions pales into insignificance, a mere shadow of what it could be."

Rene Jorgensen, a Danish NGO worker who encountered this light during an adverse drug reaction in India, tried to get across to me the sheer power of its sentient luminescence. In his scantily furnished apartment in Montréal, the main visual draw in the living room is an enormous framed print of a lighthouse surrounded by raging ocean waves. "If you stand on that ledge behind Niagara Falls," he said, "you know, where tourists can be sort of underneath or behind the water, imagine that the incredible roar and volume and energy of that water is love. It's something like that." His heightened sense of awareness, he added, was also beyond compare. "The experience is so powerful that it doesn't leave any room for doubt."

When scientists and journalists describe the "white light" that people see during near-death experiences as if they were talking about fleeting quirks in visual perception related to oxygen deprivation, they are misunderstanding the phenomenon. Yvonne Kason, for one thing, wasn't having an NDE and was not oxygen deprived. In fact, the common thread in experiences of this light is not our physical state when we encounter it, but our emotional response. "When you look at the essence of what people are going through," Kason says, referring to the variety of spiritual experiences she later dealt with as an open-minded physician, "that essence is getting closer and closer to the same thing. So if you had your experience because you're a mystic spending hours and hours in prayer, that is very similar to the yogi who is meditating, and the near-death experiencer, and the dying person having a deathbed vision. To me, these are all glimpses of the same spiritual reality."

People have abruptly encountered this light in moments of extreme depression as well as at times of acute terror. When Beverly Brodsky was twenty, she lost half the skin on her face in a traumatizing motorcycle accident. Deeply despondent, still grieving the death of her father as well, she offered up a bitter prayer to a God she didn't believe in to just let her life end. "The pain was unbearable," she wrote, "no man would ever love me; there was, for me, no reason to continue living." As she lay facedown on her bed in a rented apartment in Los Angeles, "somehow, an unexpected peace descended upon me. I found myself floating on the ceiling over the bed . . . I barely had time to realize the glorious strangeness of the situation—that I was me but not in my body—when I was joined by a radiant being bathed in a shimmering white glow." She accompanied this being "traveling a long distance upward toward the light. . . . Within it I sensed an all-pervading intelligence, wisdom, compassion, love, and truth; there was neither form nor sex to this perfect Being. It, which I shall in the future call He, in keeping with our commonly accepted syntax, contained everything, as white light contains all the colors of a rainbow when penetrating a prism. And deep within me came an instant and wondrous recognition: I was facing God."

"I've spoken with people who have had full-blown spiritual experiences," Kason told me, "with all the characteristics of near-death experiences, yet nothing happened to them. Men who fought in the wars: A grenade drops in front of them. They're absolutely certain they're going to die. 'There's a live grenade just dropped in front of me,' and boom, they go into the light and then by some freak thing the grenade didn't go off. Or another man told me about fighting in World War II, and his plane got shot down. He goes into the light, and then somehow the pilot managed to pull the plane out of spin, and they never crashed. In 1979, I didn't even know what to call what had happened to me."

Whatever the phenomenon is, it is extraordinary and transformative, and propels human beings far beyond what endorphin rushes or tricks of the optic nerve could ever achieve. It has likely driven religious conviction since the outset of human consciousness.

In the 1920s, the German theologian Rudolf Otto tried to grapple with the impossibility of understanding how light could, at the same time, be love, could also be knowledge, could swallow you—and your ego—whole. He argued that religious orthodoxy had almost fatally committed itself to defining the religious experience in rational terms. By doing so, it abandoned the very essence of spiritual experience, which is that it is *ineffable*. Indescribable. It is one culture lacking a concept for palm tree, or another society having no word for snow. The experience, until you have experienced it yourself, is totally untranslatable. That is *not* the same thing as saying "you have got to have faith because we say so." It is the opposite of that statement. Mystical light is a powerfully evident experience when you have it; you're taking nothing on faith, you are bowled over sideways. But when you have not had the experience, you are left with unconvincing words lying flat on a page.

The word "holy" denotes moral or ethical goodness. A holy man. Or just: "holy cow." The word has deteriorated in meaning, the way the word "awesome" has. ("Do you want me to put your receipt in the bag?" "Yes, please." "Okay, awesome.") For mystics and shamans, holiness refers to a radically deeper meaningfulness that rests at the very center of all religions. Otto chose to call it the "numinous" experience, borrowing from the Latin word *numen*, which roughly translates as divine power, or God's majesty.

The numinous, he wrote, "is beyond our apprehension because, in it, we come upon something 'wholly other.'" The numinous contains an element of fascination, of "strange ravishment." It contains energy, "vitality, passion, emotional temper, will, force, movement." It isn't what you saw—if you saw it—in Clint Eastwood's movie *Hereafter*, when a woman caught by the Thailand tsunami perceived vague and flickering figures in a watery white light. That doesn't begin to convey the dynamic complexity and force.

"Oh, that I could tell you what the heart feels, how it burns and is consumed inwardly!" said the sixteenth-century Italian mystic Catherine of Genoa. "Only, I find no words to express it. I can but say: Might but one drop of what I feel fall into Hell, Hell would be transformed into a paradise."

In his recent book on hallucinations, the neurologist Oliver Sacks writes that people can only hallucinate images, sounds, and odors that they are already familiar with. Yet according to people who encounter it, the numinous is as unprecedented an experience as it gets. People devote years of their lives to finding something in human society that replicates the enthrallment of their encounter with this light: they leaf through art books, visit libraries, search for the music they heard, scour art galleries. But the only place they find it, ultimately, is in how it reverberates in their souls. *It was like I'd been lost for centuries and I'd found my way home.*

This, then, is what we hear from the sensed presences, and from the dying. "I'm going home now." "I have to go home." One woman described the day after her mother died, seeing a vision of her in a hallway where her mother, leaning against a wall, said, "I am back home." Could it be that the dying and the dead are trying to tell us something that is difficult to convey? We're so busy defending the fact that we even saw or heard them at all that we can't even arrive at the point of wondering what they mean.

"Everything turns upon the character of this overpowering might," Otto wrote of the numinous experience, "which can only be suggested indirectly through the tone and content of a man's feeling-response to it." He described having a conversation with a Buddhist monk, and asking him to describe what nirvana was. "After a long pause came the single answer, low and restrained, 'bliss . . . unspeakable.' And the hushed restraint of that answer, the solemnity of his voice, demeanor, and gesture made more clear what was meant than the words themselves."

So it is with the radiant transformations we witness in the dying. There is nothing they could say that would be any more moving to us than what we see. When the psychologists Osis and Haraldsson did their deathbed vision research in the 1970s, the witnessing doctors and family members told them this repeatedly. "Patients reacted to otherworldly visions with otherworldly feelings," they wrote. "A coal miner's wife was dying of a very painful cancer. Her consciousness was clear when she said to the nurse: 'Virgin Mary! How beautiful!' The nurse said, 'She seemed to be in ecstasy—very happy.'"

When we speak of feeling peaceful, we can mean many things. When we speak of contentment or happiness, many things. An endorphin cascade in the bloodstream in response to terror is not the same as spiritual bliss; just as Romeo's love for Juliet is not the same as his love for his uncle or for Verona. The scientists who tell us how to think about this elusive yet powerful experience are specialists in their own language, not in theological texts. Only a few have tried to bridge that divide.

Andrew Newberg, a medical doctor and professor of radiology at the University of Pennsylvania, collaborated with the psychiatrist Eugene d'Aquili in the late 1990s to study peak spiritual experiences in Buddhist meditators and Franciscan nuns at prayer. The two men were curious to see if there was a neural signature to spiritual experiences, something that would distinguish them from other emotional states. When the meditators and nuns reached a point in their meditation or prayer where they felt they were attaining mystical consciousness, they were instructed (beforehand) to tug on a string, which signaled the researchers to initiate an intravenous injection of radioactive material that would show up in the brain during a

SPECT scan. After the injection, the subjects were wheeled from the room in which they had been praying or meditating to an examination room with a SPECT scanner, which whirled around them, scanning for the radioactive tracer while they lay prone on a table. (Was this distracting? Yes, it surely was, but the tracer had locked on to the peak experience, so however the poor subjects felt immediately afterwards was irrelevant in terms of science.)

The scans revealed a "sharp reduction" in blood flow to the posterior superior parietal lobe, which is the part of the brain that orients us in space. When it is damaged by a stroke, for example, it becomes impossible for a person to tell where he ends and the room around him begins. The neurologist Jill Bolte-Taylor's autobiography, *My Stroke of Insight*, recounts the results of a sudden stroke that damaged that part of her brain. Dr. Bolte-Taylor couldn't even tell her fingers from the phone as she tried to dial 911. Everything became boundaryless. We need this function of our brain to quite literally put one foot in front of the other when we walk down the street.

Experienced meditators are able to quiet or disable this brain region temporarily, preventing sensory data from reaching it and enabling them to experience what has been called the unio mystica, the sense of oneness, or the sense that Yvonne Kason had of being merged in a sea. The fourteenth-century German mystic Johannes Tauler describes the feeling as being as if the soul becomes "sunk and lost in the abyss of the Deity, and loses the consciousness of all creature distinctions. All things are gathered together in one with the divine sweetness."

As with descriptions of sentient light, this sense of dissolution into a universal whole is referenced in every spiritual tradition, from Sufism to Taoism to Christianity. "Thou shalt not love him as he is: not as a God, not as a spirit, not as a person, not as an image, but as sheer, pure One. And into this One we are to sink from nothing to nothing," said the German mystic Meister Eckhart in the fourteenth century.

"We saw evidence of a neurological process," write Newberg and d'Aquili, "that has evolved to allow us humans to transcend material existence and acknowledge and connect with a deeper, more spiritual part of ourselves." Whether that part of ourselves is within or beyond the brain, they couldn't say. "At this point in our research," Newberg wrote, "science had brought us as far as it could, and we were left with two mutually exclusive possibilities: either spiritual experience is nothing more than a neurological construct created by and contained within the brain, or the state of absolute union that the mystics describe does in fact exist and the mind has developed the capability to perceive it. Science offers no clear way to resolve this question."

Subsequent research by the Montréal neuroscientist Mario Beauregard, imaging the brains of Carmelite nuns with an MRI machine, found distinct patterns of activation in several parts of the brain, confirming Newberg's discovery that the brain is *doing something quite different* during a mystical experience than what it does when we remember, dream, or imagine. Whatever is going on, a numinous experience is not akin to other emotional or psychological states, and it isn't simply a by-product of random neural malfunction. An injury may facilitate the experience, the way a tear in the hull of the *Titanic* facilitated the inundation of seawater, but the sea, like the experience, is not a delusion.

From the point of view of the person encountering the numinous, a word like "hallucination" isn't quite up to the job of describing the sensory, emotional, moral, and psychological complexity of what is going on. "It's like the first time you jump in an ocean,"

Kason said. "You experience a lot of things at once. You experience that it's wet, you experience that it's cold, you experience that there are currents. Maybe you experience weeds, maybe you experience the salt, I mean you experience a lot of things at once that maybe later you can put words to. It was like that, there were many facets to it. The light, the love, the higher power. There was no *question* that I was . . . sort of *embraced* by a higher power. In that love, and in that light and in that intelligence, I just knew that [whatever happened] was meant to be. . . . I was in complete joy, complete love, complete contentment."

Like the meditators who remained oriented enough in this world that they were able to tug on the string and initiate the SPECT scan injection, Kason was, at one and the same time, oriented to the light and to the lake, the blizzard, the pilot and nurse also swimming, and her physical peril. "I would say part of my mind was still in the body, but most of my consciousness was not. It would shift back and forth. It's sort of like you're cooking dinner, your kid's watching TV, and you're talking on the phone, and you remain aware of all three things throughout." This description of being mindful of two worlds corresponds with what nurses and family members sometimes observe with the dying in hospice care. Nurse Maggie Callanan commented on this recently in a radio interview: "They, on some level, understand that it's odd, that they have one foot in two worlds. Do you remember the old Brownie Hawkeye cameras where, if you didn't advance the film, you'd get two pictures on one negative? Two images on one film—I think—is what the experience is like. Some of what they're saying belongs to this world, and some of it belongs to a world we cannot see."

"It was sort of like a split-screen TV, is the best I can describe it," Kason said. "The big picture was the light and the little tiny picture was the body swimming to shore, and the light was *far* more interesting." She laughed. "I still knew my body was there, but my awareness shifted to the light. At a certain point, as my body was sinking, I shifted more of my attention to it. I remember, because I was so calm, thinking, 'Oh, yes, so you do drown the third time you go down.'" Switching back into the perspective of her floundering physical form, Kason saw with her eyes at lake level that the current was carrying her swiftly toward a fallen pine tree by the shore. If she could angle herself with two more swim strokes, the current would—and did—deliver her to the tree. "And that's how I survived, because I did not have the strength to swim the last few feet to shore." The current carried her.

For another thirty minutes, Kason and her pilot, Gerry Kruschenske, languished on the edge of the lake at Devil's Elbow while the nurse, Sally Irwin, remained alive but in the water, clinging to driftwood. Their SOS signal had been caught by an overhead jet, whose pilots relayed it to the nearby town of Kenora. A complicated and heroic helicopter rescue ensued, which won the two helicopter pilots awards for valor and became the focus of media coverage. Kason kept her spiritual experience private. By the time they reached the hospital, Kason was slipping in and out of consciousness, and continued hovering above her body, severely hypothermic. The emergency room nurses covered her, at first, in a light blanket. Abruptly, she told me, "my body spoke. This is without me planning to speak. It said, 'Boy, could I use a hot bath.' Clearly, it was something higher speaking through my mouth, telling them what to do. How did my body speak this when I was not even thinking it?" Kason had no training in the treatment of hypothermia. "The nurses said, gee, maybe that would help; let's take them to the whirlpool in physio. When they put me in the bath, I felt like a genie being sucked back into a bottle. Suddenly, I'm fully back. It was like finding out the end of the story, because I didn't know how it would end until then."

It ended with life, enlightened.

CHAPTER 6

Deeper: What NDEs Tell Us About Where the Dying Go

I am wary of the sea. My sister and I once spent Christmas in a cabin on the westernmost edge of Vancouver Island, near Tofino. The power of the roaring Pacific during a tornadic winter storm tossed fifty-foot deadheads onto the beach like so many twigs. This is the ocean that David Bennett once drowned in.

In 1983, Bennett was a twenty-seven-year-old native of Syracuse, New York, working as chief naval engineer on the research vessel *Aloha*, owned by International Underwater Contractors in California. One evening in early March, the weather and the sea grew stormy. The crew of *Aloha* had just spent the day testing a remote-operated submersible in the company of its manufacturer, and now found it too rough to bring the ship safely back into port at Ventura. Bennett was asked to ferry the manufacturer and a few crewmen back to shore in a rubber Zodiac, the kind of boat that's flexible enough to take a pounding in the waves. It was already dark. Bennett scouted for the harbor from the bow, but the wind was blowing salt water directly into his face while huge Pacific swells nudged the Zodiac off course toward a sandbar a mile or so south of Ventura that was creating massive, twenty-five-foot breakers.

Before they could retreat out to sea and skirt the sandbar, a wave rose up, curled, and crashed on top of them, folding the Zodiac in half and shattering its fiberglass floor. Bennett was thrown into the gargantuan surf. "It was the most raging violent force that had ever attacked my body," he said later.

He was submerged in the ocean with no sense of which way was up, wearing an old canvas life jacket that was soaking up water like a sponge and preventing him from floating to the surface. "My years of experience as a diver had taught me not to panic, because if you panic you will surely die. I'd [also] learned that a strange thing happens to me in dangerous situations. A clear calm sense comes over me, allowing me to work through the danger. So I knew not to try to swim for the surface because I could be swimming the wrong way. I could tell by the pressure of my ears that I was in deep water, past the sandbar." He began feeling starved for oxygen. "In dive school, I had been forced to experience oxygen deprivation in a safe and controlled environment. The instructor slowly decreases the oxygen going into your dive helmet, so that as you breathe, you build up carbon dioxide." Remembering what that felt like from his training, Bennett began to realize he might die. "I couldn't hold my breath any longer, and I couldn't find my way to the surface. I clearly remember the burning, choking, and pain in my lungs as I breathed in the salt water and went through the agony of dying. [But] the agony quickly melted away into darkness."

With the darkness came silence, or what Bennett more aptly described as "a shocking absence of noise." Given what he'd just been going through, the almost deafening roar of the breakers, the frantic thrumming of his pulse, this soundlessness must have been remarkable. He noticed that he no longer felt as cold. He wondered if this was yet another stage in the experience of oxygen deprivation—one he hadn't encountered in his training. Was it a dream? It couldn't be. He knew he hadn't lost consciousness.

All at once, Bennett began to "sense a connectedness" in the dark. It was, somehow, more than mere darkness and silence; it had an emotional quality to it. How could that be? "Although I could not quite grasp the meaning of this perception, I sensed there was something going on around me." He began to become aware of a field of light, growing brighter and beginning to surround him, although he couldn't tell if he'd moved toward it—or it had moved toward him.

"As I got closer, I started *feeling* the light." This utterly unexpected sensation intensified as the luminescence grew all the more radiant. "The Light was brighter than any light I had ever experienced. As a chief engineer on a ship, I had many occasions to use an arc welder. The light emitted from the arc is so bright you have to wear protective eye gear to look at it without burning your eyes. This light was brighter than that, yet I could still view it comfortably." (This is a common remark made by people who have these experiences. Light that ought to make them squint or flinch, like sunlight on snow, is painless to gaze at.)

At this stage in Bennett's undersea adventure, having run out of air and swallowed ocean, he should have blacked out, died, or—at the very least—lapsed into an acute confusional state. Instead, he discovered that he could see more clearly than he ever had before (he could not see well without his glasses), and that, further, he could see in multiple directions simultaneously, which he found fascinating.

At some point, he realized that he was merging with the light— had become light, himself. He accepted the transformation, feeling overwhelmed and awed. "Every time I explain my experience in the light I feel a touch of the love that accompanied it. It is still incredibly emotional for me even many years later." He has reflected on the difference between this numinous experience of love and what we more typically feel: "In the light without a body I could handle that level of love because I had left the physical side of emotion behind . . . in our physical body, we feel excitement in our stomach and love can make us light-headed. In the light, I felt love, joy, passion, and excitement without the physical sensations. I had no physical reaction that might cause one to say, 'This is something I want to distance myself from.' 'I am not ready.' Or 'I am not worthy.'"

Another reaction we might have is terror at losing ourselves, at surrendering autonomy and identity in this merging. Most dystopian fiction, from Orwell to Koestler to Madeleine L'Engle, frightens readers by dwelling on the prospect of forced conformity and groupthink. To become ants in a colony, fish in the sea, drops in an ocean of light—these are alarming fates to highly individualistic Westerners. Yet in the unio mystica, the loss of self isn't experienced as frightening. On the contrary, separateness is suddenly understood to be the illusion we have been living with our whole lives.

In his survey of afterlife beliefs, *The Modern Book of the Dead*, journalist Ptolemy Tompkins described this from the Buddhist perspective: "Whether we are currently in a physical body (be it human or animal or supernatural) or in the forty-nine-day period that separates [for Buddhists] one incarnation from the next, all of us suffer from having forgotten one absolutely crucial piece of information: neither we, nor anyone else, nor any single thing or quality or situation in the entire universe, has any underlying *separate* existence whatsoever." Tompkins likens this existential amnesia to his own recurring dream where he finds himself back in high school, appalled that he faces a test and hasn't studied for it. The emotion is appropriate to the situation, the classroom is ordered, it all makes sense as a scenario, but for the fact that he hasn't *been* in high school for thirty years. That is the key missing piece of information. We can buy completely into a reality without ever realizing that we are missing one totally obvious fact, whatever that fact may be. In this case, the missing fact is that we are inseparably connected. As Sogyal Rinpoche writes in *The Tibetan Book of Living and Dying*, "The memory of your inner nature, with all its splendor and confidence, begins to return to you."

Alan Ross Hugenot, a San Francisco engineer who had an NDE during a motorcycle accident, found the experience lent him insight into why those who are dying more slowly in a hospice would seem to "know" when they were going to die. It isn't a premonition, in his opinion. "It's your decision," referring to the dying drifting back and forth between worlds. "You begin to be at peace, knowing that you're going home, and aware of that peace to come, you cease the tiring effort of holding the illusion [of separateness] together." In order to understand what the dying are demonstrating or conveying to us, it may help the inquiry to listen to those who've had NDEs.

Near-death experiences have been reported by 12 to 18 percent of the American population. They differ from what happened to Yvonne Kason insofar as they occur during periods of *physical* near death, which is to say, during incidents of drowning, cardiac arrest, and coma. Over 90 percent of cardiac arrest survivors have no memory of what happened before and after their resuscitation due to brain trauma, so the true incidence rate of NDEs remains unknown. An interesting example of this is what happened to forty-two-year-old Vancouver Realtor Tony Cikes, who doesn't remember that his heart stopped for two minutes during a routine operation in 2011. But when he roused to consciousness, "I knew that something had saved my life. It was a bizarre feeling. I felt off and I couldn't quite explain it. I felt that someone had saved my life even though I didn't know what had happened. I had no idea that my heart stopped." Yet, upon awakening, he discovered this surge of compassion, which he could only figure out how to describe as an energy. The nurse who was attending him in recovery was the first recipient. "I wanted to help her. It was an awakening to being compassionate. I felt this love inside me. The doctors were surprised."

The psychologist Carlos Alvarado, who has written extensively on the subject of NDEs, fixes the known incidence rate at 17 percent. To decide what to include as a bona fide NDE, he used the Near-Death Experience Scale developed by University of Virginia psychiatrist Bruce Greyson, head of the University's Division of Perceptual Studies and former longtime editor of the *Journal of Near-Death Studies*. This scale comprises sixteen items, including feelings of peace, being surrounded by light, experiencing cosmic unity, seeming to enter another world or landscape, encountering other beings, undergoing a life review, and receiving new insight or knowledge. A classic NDE is considered to have occurred when more than seven of the sixteen items are involved. The greater the number of items, the deeper the NDE. David Bennett, submerged in the ocean, had all of them.

After Bennett merged with the light, he encountered what he could only roughly describe as silhouettes or presences that projected "waves of love and compassion." Although he didn't recognize anyone in particular, he experienced these presences as family, or as more familiar to him on some level than the people he was related to in his day-to-day life. A soul family. "I just knew unequivocally that I was home," he said, echoing others, "and it felt so exceptionally magnificent."

In the company of these presences, Bennett proceeded to review his life, during which "Time doesn't exist," he said. "It's one of the things I have a hard time grasping. I haven't come across anything [in literature] that adequately describes the sense of timelessness. The detail [of my life] was richer and more vivid than in this life. It was my interactions, and others' as well. We're unbounded. I could envision multiple streams of experience."

He was, at times, embarrassed by what the presences who encircled him were witnessing him do in his life review: the toes he stepped on to become a chief engineer at such a young age, the way he treated certain people, his rough upbringing in foster care. But they exuded no judgment. "Their form of communication involved projecting a knowing and comforting energy that contained more information in a millisecond than our [typical] thoughts could assemble in a day." The most captivating part of his life review, he said, was gaining the perspective of those his actions had affected. He gave me the example of a bar he scarcely remembered being in, where he had picked a fight with an unwilling foe, goading and taunting the man—he'd been in his fair share of brawls—but now he observed and felt the emotional energy his aggression had generated, the sourness, like an acrid smell.

Bennett was lucky, in a sense, that all he'd done by the age of twenty-seven was throw some punches, steam ahead in his career, and occasionally behave like a jerk. One man, a Californian arms dealer who had sold weapons in Latin America, had a near-death experience when he was struck by lightning in 1984, and underwent a life review where he felt the ripples of pain in every family who'd lost someone murdered by one of his weapons. After his NDE, he became a hospice volunteer.

Small kindnesses also appear in the life review experience, driving home the point that love and compassion are "the absolute coin of the realm" as the neurosurgeon Eben Alexander, who had an NDE in 2008, says. Bennett hadn't even remembered his interactions with one woman who showed up in his life review. "When I was young," he says, "I was brash. At one point I was working in a butcher shop, and I took a liking to this lady—and she was the grumpiest lady around. But for some reason, I took a shine to her. For months I did special things for her." Mostly, he was trying to coax a smile out of the woman, just to see if he could, a little game. "But I got to reenvision it, and see how all those small efforts I made affected her."

After a million years or none, twenty-seven years or five seconds, Bennett suddenly "heard a clear, distinct voice that didn't emanate from my [presences]. I sensed it came from the Light itself. I stopped paying attention to my life experiences and the group. I listened carefully to an incredibly loving voice, which told me, 'This is not your time; you have to return.'"

He had, by now, been deep underwater for several minutes. Bennett's arm had gotten entangled with a bowline dangling from the broken Zodiac on the sea's roiling surface. As a wave shoved the boat across the surface, the still-attached line jerked Bennett upward, dislocating his shoulder and thumb while dragging him to the water's surface. He watched this from an out-of-body perspective and was still observing when another wave flung him against the pontoon with such force that it expelled water from his lungs, much the way you might get the wind knocked out of you. At this point, he "experienced a rushing, buzzing vibration," and found himself roused to consciousness within his cold, dense-feeling, injured, and gasping body. The shock was physical and emotional. "Separating from the Light and rejoining my body was the hardest thing I had ever been asked to do. It was more painful than drowning."

Bennett's fellow travelers regrouped with him around the shattered vessel—all had survived—and now they collectively began to kick-swim to shore. When they made it, they were too shocked and exhausted to talk to each other, first lying on the sand ebbed of strength, then walking to the highway to flag down a lift back to the marina and to Bennett's van. (One crew member popped Bennett's shoulder back in for him.) Speechless, they drove back to Santa Barbara. By this time, it was after midnight. Bennett's wife, unbeknownst to him, had just woken from a nightmare in which he had died. When he arrived at their apartment bedraggled and covered in sand, she deduced what must have happened. A trained nurse, she set about easing him into a warm bath.

But also, "She was trying to question me, get a reaction. I wasn't responsive because I couldn't stop thinking about my experience. Eventually I said, 'Hon, I think I died tonight.' This confession so frightened her that she started slapping me. Afterward she told me that she hit me because she thought I was in shock." But Bennett interpreted the gesture as an act of censure, like Pandora trying to slam down the box lid, fearful of what was within.

"I shut up for eleven years," he says. "It was too much for me, it was overwhelming, I couldn't face it." This statement brought to mind, for me, another bit of T. S. Eliot's poetry: "We have lingered in the chambers of the sea, by sea-girls wreathed in seaweed, red and brown. Till human voices wake us, and we drown."

Those human voices are clamourous in their efforts to offer explanations for the NDE. A number of scientific theories have been offered since they were first formally described by the psychiatrist Raymond Moody in his 1975 book, *Life After Life*. NDEs have been attributed to psychopathology, altered blood gases, sleep disorders, and changes in brain activity around death. Every theory, no matter how tentative, invariably gets headlined in the media as "NDEs explained." As I write this, the *Guardian* newspaper arrives in my mail featuring the headline: "Near-Death Experiences: The Brain's Last Hurrah," citing a new study done at the University of Michigan where induced heart attacks in rats demonstrated continuing electrical activity in the rodent brain for thirty seconds postmortem.

The small number of scientists who engage in NDE research in depth, interviewing people who have had them, tend to be uncomfortable with sweeping conclusions. The power and complexity of NDEs have not been explained to their satisfaction. Mysteries about human consciousness abound.

"The most important objection to the adequacy of all [these] theories," writes the psychiatrist Bruce Greyson, "is that mental clarity, vivid sensory imagery, a memory of the experience, and a conviction that the experience seemed 'more real than real' are the norm for NDEs." Yet they occur in brain states that should not have any kind of coherent thought process at all, regardless of whether you want to call that process a dream, a memory, or an elaborate hallucination. Science has yet to explain how that could be. Severe brain impairment due to oxygen deprivation or powerful sedation doesn't set up the right conditions for incisive thought.

Jayne Smith, a retired actress from Delaware, had an NDE before they had a name, in 1952. "I thought I was the only person in the entire world that this had ever happened to," she said. "I couldn't imagine why." Attractive, crisply dressed, she spoke very matter-of-factly. The NDE took place during the birth of her second child. "They used—back in those ancient days—a device called a trilene mask, which was a bracelet on the wrist with a nose cone. When your labor pains became more than you wanted to cope with, you would put it to your nose, breathe in, and for a few seconds lose consciousness. I remember, after a particularly disastrous pain, putting that nose cone up to my nose and inhaling as deeply as I possibly could. When that inhalation was over," she had overdosed on general anesthetic. She went into cardiac arrest, and the doctors began emergency resuscitation.

But that's not what she experienced. "I felt—I felt—myself rising up out of my body, up through the top of my head, and was in total blackness. I didn't see a tunnel. I thought, this is not right, I should be unconscious, but I'm still awake. Something is a little bit off; I don't know what it is. And then, as quickly as that, I found myself standing in a kind of gray mist, and I knew right away what had happened."

Here, Smith's voice took on a musical sense of wonder as she recalled the moment: "I remember distinctly thinking, 'I know what it is, *I've died*. I've died.' And I was so overcome with joy, because I had not been annihilated. I thought, 'I am still me, I am still here,' and I felt so grateful."

What is striking about NDEs is how the thread of consciousness continues, featuring a coherent sequence of thoughts, from the person being awake to finding themselves in an altered state. This is not what happens when we slip from wakefulness into dreams. When we fall asleep, whatever it was we were thinking about beforehand drifts away. But Smith knew that she was in labor in a delivery room, self-administering gas, and then she questioned why she no longer felt pain but wasn't unconscious, and from there the conscious awareness continued, the way it had for David Bennett: "I remember joy pouring out of my being," she said, "and as I did that, the gray mist dissipated and everything became brilliant white light."

At the time of her experience, in the early 1950s, doctors would have assumed that she was having a hallucination brought on by her anesthesia—had she told them about what happened, which she didn't. As it stands, about 2 percent of patients hallucinate briefly before surfacing from general anesthetic, but the content of their hallucinations is disjointed and random. Unconsciousness during anesthesia, as measured by EEG, is associated with an immense quieting of brain activity.

Smith's rapturous response to the love surrounding her grew so intense that she worried she would shatter. As if sensing this, the light dimmed a little, and then revealed or unveiled an exquisite meadow featuring colors that she had never seen before. "The colors were extraordinary," she said. On a ridge above the meadow she saw some figures, to whom she went and wordlessly communicated. "There was a block of knowledge," she continued, touching—again—on an element that appears frequently in deeper NDEs, "or what I later suspected was a *field* of some kind, that came in and settled on me all of a piece, and I suddenly knew that I was eternal, that I was indestructible, that I always existed and I always would exist. There was no *end*, there was no end."

Smith recovered, delivered a healthy daughter, and said nothing about her NDE for twenty-five years. Although by the time I heard her speak about it, more than fifty years had passed, her facial expression during the telling was gradually transfigured by joy. Hers was the first story I heard after my father and sister died, and I felt so consoled that, when I later ran into her at a meeting in Washington in 2013, I actually hugged her. (She reciprocated in amused surprise.) I was only disappointed that she could say nothing of meeting a sister who'd gone into death before her.

The closer that NDEers are to actual death—such as being clinically brain-dead or in cardiac arrest—the more likely they are to encounter deceased people in the NDE, according to research by the University of Virginia psychiatrist Emily Williams Kelly. This is reminiscent of the appearances that the deceased make in deathbed visions. In fact, both phenomena can occur to dying people, as the terminally ill may have NDEs in palliative care days or weeks before they die. In some cases, the person perceived is not known to be dead yet by the person who has the NDE. So the content of an NDE cannot be attributed to expectation. For instance, a nine-year-old boy with meningitis awakened from a coma and told his parents that he had seen his teenaged sister on the other side—she had in fact just died in a car accident, unbeknownst to the boy and his parents. Children, in particular, will sometimes spontaneously recognize photographs of dead relatives in the family album after their NDE, as detailed in several studies. The Dutch cardiologist Pim van Lommel reports that people who had deep NDEs—featuring more of the elements on the Greyson scale, from out-of-body perceptions to sightings of deceased loved ones and encounters with the light—were more likely to die within thirty days of their cardiac arrest than his other patients. Yet, "medically they were no different from the other patients. I cannot offer an adequate explanation for this." (Oddly enough, I was going to interview someone who had had a deep NDE on a cardiac ward in Vancouver, and although he was young, with a good recovery projected, he died suddenly.)

About 56 percent of Westerners who have NDEs experience themselves going out of body. A child who sustained a major head injury in a car accident spent months in a coma before returning to waking consciousness with slurred speech and a permanent mental disability. Nonetheless, he recalled his time in the hospital this way: "I felt like the inner part of me, I felt like my inner part in me was up like a ghost and I felt like my inner body was taken out of me. I felt like I was going out, like my inner body was going out of me and I felt like, I felt like a dummy almost, that my body was like a dummy and I was outside it." (Two years after the accident, when he visited the ICU he'd originally been treated in before being wheeled elsewhere for the sojourn of his coma, he went straight to the correct section and specific bed.)

Other features of the near-death experience are more varied. People in indigenous cultures, for instance, have never reported going through anything resembling a tunnel. Instead, they journey across otherworldly landscapes, much the way their shamans do. Here is a Maori account: "I became so ill that my spirit actually passed out of my body. My family believed I was dead because my breathing stopped. They took me to the marae [a communal sacred place in Polynesian cultures], laid out my body, and began to call people for the tangi. Meanwhile, in my spirit, I had hovered over my head, then left the room and traveled northwards, towards the Tail of the Fish." She sailed over rivers and mountains "until at last I came to Te Rerenga Wairua, the leaping-off place of spirits." There are accounts in Hawaii of walking toward Pele's Pit, a volcano into which dead souls hurl themselves. An Australian aboriginal described an NDE in which he set off by canoe until he reached an island where he met traditional spirits and dead relations. Journeys on land and by water are likewise taken in Africa and South America. While the cultural shadings differ, what all the experiences had in common was the person being brought to a halt at some point by some relative or spirit, and they were told to go back, that it wasn't their time.

Beyond that, the indigenous accounts are uniform in *what they lack*. There is no account collected from a hunter-gatherer, "primitive cultivator," or "herder culture" that features the kind of life review that David Bennett described. In these people's cosmological worlds, a sense of unity and interconnectedness is already fully expressed in day-to-day life. Perhaps to them, the Western view of an individual fate (and personal accountability) is less relevant. As the sociologist Allan Kellehear notes, "Individuals are no more responsible than the world."

Accounts from India, China, Thailand, Tibet, North America, and Europe do feature the experience of reliving and learning from some (or all) of their actions. A 2013 study of severely brain-injured patients suffering posttraumatic amnesia in Guangzhou, China, uncovered three NDEs out of eighty-six patients; one had a life review that involved a "structure of many animals," in which the impossible creature "reminded him of the wrong deeds he had done in his life" by "sending him light rays that pierced his chest region and created huge waves of negative feelings like fear, sadness, and anger." In India, people often encounter the spiritual equivalent of irritated bureaucrats, telling them they're the wrong one, that the *Yam Dhut* (death messenger) accompanying them has brought the wrong person. The *Yam Dhut* also shows up in the hospital room in deathbed visions in India. "Clearly, NDErs meet an assortment of social beings, and their previous experiences shape *their interpretation* of the identity, function, and meaning of these beings," Kellehear wrote. In Carl Sagan's science-fiction novel *Contact*, Dr. Ellie Arroway travels to meet some unknown alien intelligence that has broadcast a signal to Earth and finds herself on a beach that resembles one from her Florida childhood. The alien she then encounters projects himself as her beloved, deceased father so as not to frighten her.

Perhaps we encounter what we can emotionally and conceptually relate to. People remember the emotional content of the NDE with extreme clarity, regardless of its symbolic or cultural content. The neurologist Steven Laureys of the Coma Science Group at the University Hospital of Liège, Belgium, recently tested the hypothesis that people who have NDEs merely imagine them. The researchers compared the memories of three groups of coma survivors:

those who had had NDEs; those with memories of their coma without NDEs, such as recalling being talked to by a family member; patients with no memories of their coma; and a control group who had never been in a coma. Each group was asked to recall five types of past events: recent and old memories, such as a first kiss; recent and old imagined event memories, such as a daydream; and one target memory—the NDE, for those who had one; the memories of being talked to during coma, for those who had that experience; and, for the control group, a memory from childhood.

The researchers discovered that memories of a near-death experience were by far the sharpest, clearest, and most detailed. NDEs "seem to be unique, unrivalled memories." They were not imagined. Intense emotion sears memories into our brains. Heartbreak, the birth of a child, being caught in a hurricane: people recall these in greater detail than the daily plod to work. Neurologists call them flashbulb memories, in that a "highly emotional, personally important, and surprising event can benefit from a preferential encoding that makes them more detailed and longer-lasting." NDEs appear to trigger the most preferential coding of all.

There is no evidence that NDEs become confabulated over time. Even children who have experienced them carry forward precise recollections across the years. "Interestingly," the Belgian researchers noted, "NDE memories in this study [also] contained more characteristics than coma memories, suggesting that what makes the NDEs 'unique' is not being 'near-death' but rather the perception of the experience itself." This was Rudolf Otto's point about encountering the holy. God, spiritual majesty, the Ground of Being, whatever you choose to call it—*uniquely elicits* this emotional response.

The researchers concluded that the only explanation, at least from a materialist point of view, is that NDEs are elaborate hallucinations, because hallucinations can be perceived—and responded to—as if they are real. While receiving her doctorate from Cardiff University, in Wales, the critical-care nurse Penny Sartori compared NDE narratives with medication-related hallucinations, however, and found distinctive differences, as has also been noted with death-bed visions. Hallucinations related to morphine and other palliative drugs "tended to be random and non-specific," she said. (Sartori also measured blood gases during her five-year study of NDEs and did not find any correlation between reduced oxygen levels and which patient had the experience.)

As Andrew Newberg and Eugene d'Aquili have pointed out, "when hallucinating individuals return to normal consciousness, they immediately recognize the fragmented and dreamlike nature of their hallucinatory interlude, and understand that it was all a mistake of the mind." This is not the case for those who encounter the numinous. Wrote Teresa of Ávila in sixteenth-century Spain: "God visits the soul in a way that prevents it doubting when it comes to itself that it has been in God and God in it, and so firmly is it convinced of this truth that, though years may pass before the state recurs, the soul can never forget it, never doubt its reality."

A near-death experience, a mystical, numinous experience, actually feels more real than our present, material reality. Here is what the psychologist Carl Jung wrote, after his near-death experience during a heart attack in Switzerland in 1944: "The view of city and mountains from my sickbed seemed to me like a painted curtain with black holes in it, or a tattered sheet of newspaper full of photographs that meant nothing. Disappointed, I thought, 'Now I must return to the "box system" again.' For it seemed to me as if behind the horizon of the cosmos a three-dimensional world had been artificially built up, in which each person sat by himself in a little box.

And now I should have to convince myself all over again that this was important!"

In the aftermath of their NDEs, people describe the world around them as flat, dull, two-dimensional, filled with stick figures and "paper cutout dolls." Asked simple "agree" or "disagree" questions, 56 percent agree that the NDE realms were "1,000 times stronger" as a reality than their everyday reality.

While the grieving populace may feel abashed and self-doubting about their telepathic impressions and encounters with sensed presences, people who have NDEs are often driven by the overwhelming intensity of their experiences to make dramatic changes in their lives, which are absolutely turned upside down. They may break with their churches, their professions, sometimes their spouses. Here they were, perfectly ordinary people who'd been going about their business, whatever the business of their lives happened to be: a chemist working in a lab, a teenager on a road trip, a Vietnam vet being treated for lung cancer. All of them unsuspecting and then—bam—their world radically and irrevocably altered by an intimation of God.

For several days after his drowning, David Bennett went about in total disarray. He had developed a kind of synesthesia, the unusual perceptual condition in which one's senses are crossed, so that you can hear words or see numbers as colors, for instance. For Bennett, the effect was completely confusing. "I could see the life energy in my surroundings. There was an aura of light around all the plants and rocks in the planting beds. I could feel and touch them without physically touching them with my hands. My engineer brain kept trying to figure it out. How is this possible? I could hear rhythms, notes."

The experience of heightened perception and blended senses is common for people who have had NDEs and other numinous experiences. The New York journalist Maureen Seaberg, who has had synesthesia all her life, experiencing the alphabet as a sequence of luminescent colors, calls it "a God hangover." Although Seaberg herself was born with synesthesia, as was the violinist Itzhak Perlman, and possibly the novelist Vladimir Nabokov, she senses something transcendent in the capacity, and has found references to it in ancient religious texts. The Hebrew Bible, in some translations for example, says "All the people saw the voices" when Moses was on Mt. Sinai. Another scriptural passage describes the conferring of instantaneous knowledge that Jayne Smith described in her NDE. From the book of Acts:

"When the day of Pentecost had come, they were all together in one place. And suddenly from heaven there came a sound like the rush of a violent wind, and it filled the entire house where they were sitting. Divided tongues, as of fire, appeared among them, and a tongue rested on each of them. All of them were filled with the Holy Spirit and began to speak in other languages, as the Spirit gave them ability. . . . All were amazed and perplexed, saying to one another, 'What does this mean?'"

The journalist Maria Coffey unearthed an intriguing account of people hearing messages in languages they didn't actually know. In 1985, the Spanish mountaineer Carlos Carsolio "was climbing Nanga Parbat with a Polish team, attempting the south spur of the Rupal Face, the biggest mountain wall in the world. Conditions were dreadful, with blizzards so dense that often the men lost sight of each other. Through the ascent they communicated by two-way radio, and their conversations were recorded by people at Base Camp. Before they reached the top of the face, their food and fuel ran out. During their desperate descent, they all felt very close to death.

"Back at Base Camp, as they recovered from the ordeal, they listened to the recordings of their conversations. They were shocked. The Polish climbers spoke no Spanish and Carsolio spoke no Polish, so normally they conversed in English. During the final part of their ascent, however, and on the way down the mountain, they had all been speaking in their native languages. Listening to the tapes, the Poles couldn't understand Carsolio and he couldn't understand them.

"'But when we were up there we had understood each other perfectly,' says Carsolio. 'We had opened some channels, to another level of communication.'"

The thirteenth-century Jewish mystic Abraham Abulafia engaged in prayers that caused the letters of the alphabet to come across to him as musical notes, a practice of enhancing and blending perceptual awareness that he called "breaking the seal of the soul." Research at the University of California, Irvine, by the psychiatrist Roger Walsh determined that 86 percent of advanced meditation practitioners experienced synesthesia, whereas only 35 percent of the less-experienced meditators did. A very small percentage of the general population experienced it. Said David Bennett, of his time beneath the sea: "My perceptions were unbounded."

Perhaps something about the numinous experience itself, whether cultivated by prayer and meditation or abruptly imposed by terror and peril, breaks the seal of bounded, embodied senses. A study of people who had near-death experiences by the Australian sociologist Cherie Sutherland found that 83 percent had these kinds of synesthesic aftereffects, including enhanced or extrasensory perceptions such as telepathy and clairvoyance. Meanwhile, Brown University's Willoughby Britton discovered that people showed differences in their left temporal lobe activity after an NDE, suggesting that the experience had somehow altered the neural circuitry. Synesthesia has been associated with increased neural "hyperconnectivity" in the temporal lobe. So, here we have hints of how the brain may be accommodating (or generating, depending on your point of view) an experience of expanded consciousness.

Most people "were surprised and disrupted" by these changes to their perception of the world. A Toronto architect who experienced herself go out of body when she almost hemorrhaged to death after the birth of her son in 1990 told me that, in the aftermath, humanity felt to her like a clanging bell. "I would walk down the street and be hit by all these emotions, how everyone passing me was feeling, I could feel it—wham-wham-wham. It was so difficult that I avoided being in social gatherings. Six months later, I remember being in a crowded room for a business function and it was almost *annihilating*." She didn't connect this feeling of being overwhelmed to her out-of-body experience, but thought it might have been part of some kind of postpartum crisis. A Japanese-American businessman who had an NDE during a motorcycle crash became able to empathetically sense others' emotional pain, which overwhelmed him in the middle of sales meetings.

Alan Ross Hugenot, who had an NDE during a motorcycle accident, told me, "We become outsized when our consciousness leaves our body during the NDE, and then I don't think it entirely fits back in when we return." It's as if all the filters that the brain evolved in order to screen out distracting or irrelevant signals in the world around us no longer suffice. "I've had to develop strategies to guard myself against other people's emotions." To demonstrate, he raised his bent arm at me, like someone fending off a vampire.

Yvonne Kason knew a friend had meningitis before the friend was aware of it, because it flared up as a vision in front of her while she was driving over to the friend's house. The perception—an anatomical image of an inflamed brain—was utterly disconcerting. When Kason arrived, her friend had a headache, and Kason, feeling a bit crazy, warned her to monitor it closely and pay attention if her neck felt stiff. Within the day, the friend was admitted to the hospital for treatment.

Occasionally, someone returns from the near-death experience with a previously unknown talent. An orthopedic surgeon in New York State, Tony Cicoria, suddenly gained an obsession with learning to play and compose for the piano after he was struck by lightning while standing in a phone booth in the early nineties and went into cardiac arrest. He would go on to give several concerts. When I met him in 2013, he said that he was still trying to understand what his musical gift was intended for, in terms of healing others. He had no sense that the music was for him, alone, to practice and enjoy. That hadn't even crossed his mind.

Learning to live with a reversal of perspective on what matters can be extremely challenging. In one study, over half the people who had had NDEs were found to have "moderate to major" problems taking up the reins of their lives in the aftermath. Seventeen percent considered suicide in order to return, and one actually bit through her breathing tube moments after her NDE in a Dutch hospital, plunging herself back into cardiac arrest. Seventy percent felt isolation and depression, initially, for not being able to share what they felt they now knew about a parallel or higher reality. "It's sad that you can't talk freely about it," one woman said. "I feel penned in. I have so much exuberance and no one to hear me." Like the dying and the dead, they have something to say of great import.

They are also caught, in a sense, between worlds. David Bennett told me, without the faintest trace of uncertainty: "When life is over, you know you're going home. And that creates a longing. It's emotionally difficult." What an extraordinary statement for a nondepressed person to make! An optimistic yearning for the grave? People who have experienced an NDE become, in effect, homesick for that numinous reality they glimpsed. Lamented the Spanish mystic Saint John of the Cross: "I no longer live within myself / and I cannot live without God / for having neither him nor myself / what will life be? / It will be a thousand deaths / longing for my true life / and dying because I do not die."

Military veterans who have survived blast injuries—increasingly common in the wars in Iraq and Afghanistan—are radically challenged by the contrast between their NDEs and their ordinary realities. They find themselves going from the hunkered-down hell of battle to a realm of joy and peace—and abruptly back again into the military conflict. One platoon commander described being unable to discipline his troops after his NDE. "All I wanted to do is put my arm around them and say: 'It'll be okay.' The army's not big on hugging for discipline. Trust me." He had to resign.

Many vets can't talk about their experiences at all, because they will be assessed as psychiatric risks, according to retired Colonel Diane Corcoran, who began counseling soldiers with NDEs like the platoon commander I just mentioned when she was a young nurse in Vietnam. Admit that you've spoken with a being of light, and you will instantly be diagnosed as psychotic and either heavily sedated or discharged. After the World Wars, a great and probing literature emerged about what men weren't allowed to say of their horror, how they couldn't express terror, guilt, and anxiety, how their experiences were carefully mediated in the sanatoria and presented to the public as "shell shock." A commensurate literature has yet to arise around the modern experience of the spiritual in combat. Soldiers who have psychic intuitions, sense presences, or have NDEs are all understood to be suffering from PTSD.

An unexpected aspect of combat NDEs is that, because soldiers are often injured together in a blast, they will sometimes go out of body *together* and witness one another in astral states, Corcoran has learned. There is a variation of this phenomenon called the shared death experience, which is not unlike the shared illness experience in telepathic impression cases. Researchers of it are not talking about all parties actually dying and coming back, but rather some sort of shared or entangled perception between people. Shared death experience has become a new focus of research by Raymond Moody, the psychiatrist who founded the near-death experience field. Dr. Moody feels that shared death experiences change the conversation about what NDEs are. The psychologist Joan Borysenko described having such an experience when her eighty-one-year-old mother died in the Beth Israel Deaconess Medical Center in Boston while Borysenko was on the faculty at Harvard. The room seemed to fill with a brilliant light; both she and her teen son witnessed her mother rise spectrally—and unexpectedly—out of her body. Raymond Moody had rarely heard of shared death experience cases when he first began collecting stories of NDEs. For decades in the second half of the twentieth century, doctors and nurses were more likely than family members to be present at the moment of death, but after custom shifted in favor of family vigils, he began to hear about these curious experiences from ordinary people.

Here is an example: one woman, whom Moody doesn't identify, was holding her son's hand when he died. She "felt the life force surge from him, somewhat like an electric current, although vibration might be a more appropriate term." The room seemed to yield to a sea of light, and she saw visions of her son's life. Although she realized it would sound crazy to other people, she told Moody, "I felt the joy of his release."

In the course of his research, Moody noticed several common elements to the shared death experience, although not all of them feature in every case. They include a perception of changed geometry, which is bizarre. "This trait is difficult to describe," Moody writes, "because it takes so many different shapes. It is also one that is not found in near-death experiences." Rooms expand, collapse, dissolve.

Another frequent perception is the sound of exquisite music. Some, like the soldiers who have spoken to Colonel Corcoran, share an out-of-body perception. For example, this account from a woman in Virginia: "The night Jim died, I was sitting next to him, holding his hand, when we both left our bodies and began to fly through the air. It was amazing, frightening and puzzling. Above us was a bright light and we were headed directly for it." At this point, the woman was "sucked back" into her body and found her husband dead.

About 50 percent of people who have NDEs move beyond an encounter with light and also perceive places—cities or gardens. Sometimes, Moody has discovered, this also happens in a shared death experience. Here is an account from a woman whose close friend had just died as she sat with her:

"I was suddenly walking up a hill with Martha and we were surrounded by light. Not an ordinary light, but everything around us—plants, the ground, even the sky, glowed with its own light."

Also present in the shared death experience, according to Moody, is a border of some kind, beyond which the living cannot follow their beloved. It has been seen as a bridge, a river, a tree, or a verbal command to go back.

These experiences, whatever causes them, require a different set of scientific explanations from those that attempt to account for

NDEs. It may be that in the highly charged atmosphere of the dying person's room, our empathy and yearning lead us into a form of shared consciousness. In 2013, the late film critic Roger Ebert wrote about his wife's intimation that he was still vital and present during an episode of cardiac arrest. Chaz Ebert appears to have a talent or capacity for these heightened intuitions. In *Salon* Roger Ebert wrote, "After the first ruptured artery, the doctors thought I was finished. My wife, Chaz, said she sensed that I was still alive and was communicating to her that I wasn't finished yet. She said our hearts were beating in unison, although my heartbeat couldn't be discovered. She told the doctors I was alive, they did what doctors do, and here I am, alive.

"Do I believe her? Absolutely. I believe her literally—not symbolically, figuratively, or spiritually. I believe she was actually aware of my call and that she sensed my heartbeat. I believe she did it in the real, physical world I have described, the one that I share with my wristwatch. I see no reason why such communication could not take place. I'm not talking about telepathy, psychic phenomenon or a miracle. The only miracle is that she was there when it happened, as she was for many long days and nights. I'm talking about her standing there and knowing something. Haven't many of us experienced that? Come on, haven't you? What goes on happens at a level not accessible to scientists, theologians, mystics, physicists, philosophers or psychiatrists. It's a human kind of a thing."

Some NDEs can be connected directly to a period of flat-lined brain activity by observations made afterward by patients themselves. The cardiologist Pim van Lommel describes a case based on his Dutch study in which a cardiac arrest victim was found lying comatose and cyanotic in a field thirty minutes before his arrival at the ER; yet after his recovery he was able to describe accurately the circumstances of his resuscitation. Skeptics have argued that these "observations" are false memories, conjured out of medical dramas like *House* or *Grey's Anatomy*. To test this hypothesis, the American cardiologist Michael Sabom interviewed patients who reported NDEs, and also interviewed cardiac patients who had *not* had NDEs, asking each group to describe their cardiac resuscitation procedure as if they'd been watching it. Eighty percent of those who had no memory of an NDE, Sabom found, made at least one major error in their descriptions, whereas none of the NDE patients made errors. His findings were replicated by doctoral researcher Penny Sartori in 2008 with a study of hospitalized intensive care patients: those who reported leaving their bodies during cardiac arrests described their resuscitations accurately, whereas every cardiac arrest survivor who had not reported an out-of-body experience (OBE) described incorrect equipment or procedures.

In addition, Sabom found that the NDE group "related accurate details of idiosyncratic or unexpected events during their resuscitations." Said one: "I was above myself looking down. They was [*sic*] working on me trying to bring me back. 'Cause I didn't realize at first that it was my body. I didn't think I was dead. It was an unusual feeling. I could see them working on me and then I realized it was me they were working on. I felt no pain whatsoever and it was a most peaceful feeling. Death is nothing to be afraid of. I didn't feel nothing. They gave me a shot in the groin." The shot was a needle drawing arterial blood from his left femoral artery to test oxygen levels. (The oxygen levels were above normal.)

In another example from research by the University of Virginia psychiatrist Bruce Greyson, a patient described leaving his body and watching the cardiac surgeon "flapping his arms as if trying to fly." The surgeon verified this detail by explaining that after scrubbing, to keep his hands from possibly becoming contaminated before beginning surgery, he had developed the unusual habit of flattening his hands against his chest, while giving instructions by pointing with his elbows.

In a recent review of ninety-three published reports of potentially verifiable OBE perceptions, 43 percent were found to have been corroborated to the investigator by an independent informant (usually by nurses and doctors). Of these, 88 percent were completely accurate. "The OBE phenomenon is interesting," says the science journalist Jeff Warren, "because you get these incredibly detailed perceptions happening in people with apparently no brain activity. I wouldn't say this is conclusive proof of the existence of mind beyond body—it's foolish to conclude anything about reality—but if you *are* in the game of proving and disproving, it's definitely a piece of evidence worth taking seriously."

The people who come across NDEs most often are medical professionals, which is why so much of the near-death research has been initiated by cardiologists, anesthesiologists, critical-care nurses, and surgeons who have been confounded and intrigued by what they see in their operating rooms and wards. Sam Parnia, director of cardiopulmonary resuscitation research at SUNY in Stony Brook, New York, began investigating NDEs in the context of cardiac resuscitation, and soon heard accounts from his own medical colleagues. Here is the British cardiologist Richard Mansfield, from the very early days of improved resuscitation methods in the 1960s, recalling to Parnia a patient who had died and been left, expired, on the table for fifteen minutes. Mansfield went back into the room to check how many vials of adrenaline he'd administered in his futile resuscitation effort, so he could finish writing the death report. He discovered that the man had spontaneously revived and could describe details of what had just happened.

"He told me everything that I had said and done, such as checking the pulse, deciding to stop resuscitation, going out of the room, coming back later, looking across at him, going over and rechecking his pulse and then restarting resuscitation. He got all the details right, which was impossible because not only had he been asystole and had no pulse throughout the arrest, but he wasn't even being resuscitated for about fifteen minutes afterwards. What he told me really freaked me out." Parnia, in turn, also felt something along the lines of being freaked out. Since then, he has heard numerous first-hand accounts of this kind from startled physicians.

Dr. Tom Aufderheide, for instance, told Parnia and other attendees at a 2012 conference on cardiovascular care that he had a patient in cardiac arrest—it was actually his first patient in cardiac arrest as a brand-new doctor—who kept being revived and then rearresting over the course of eight hours. After an initial flurry of code blue attention, the more senior doctors eventually left the continuous resuscitation efforts to Aufderheide, who was furious at them and thought: *"How could you do this to me?"*

When the patient's lunch arrived, some triangulated sandwich or tepid soup, the young doctor decided to eat it. "I was hungry. So I ate his lunch. I certainly couldn't leave his room, and he wasn't going to eat it." The patient finally stabilized several hours later, and remained in the hospital for a month. Before he left for home, he spoke to the young Dr. Aufderheide of having had an NDE—the light, the tunnel, all the shattering extraordinariness of it—and then he added teasingly, "You know, I thought it was awfully funny . . . here I was dying in front of you, and you were thinking to yourself, *'How could you do this to me?'* And then you ate my lunch." (This story has shades of that of the nineteenth-century Alpine climber who remonstrated with his guide for drinking his Madeira and eating his chicken.)

"The one common feature among all these accounts," Parnia later wrote, "was that patients with cardiac arrests had come back and recalled incredibly detailed accounts of conversations and events relating to the period when they were seemingly dead to their physicians. Specifically, they all claimed to have been able to see the events relating to their own cardiac arrests while watching from a point above at the ceiling."

None of this would surprise Tibetan monks, who sit beside the deceased and read aloud from the Book of the Dead because they assume the words can be heard. Nor would it have shocked our ancient kin, who dreamed up ways to assist revenants in finding their way from this realm to another. But two things happened in the interim. First, death was medicalized and secularized, removed from sacred context. And then, as a subsequent and unexpected development, patients began returning from it. Dr. Richard Mansfield's exchange happened at the dawn of the new era of modern medical resuscitation in the 1960s, before which there would have been far fewer NDEs to report in the medical setting. Since then, playing God and raising Lazarus has become an increasingly widespread medical intervention, which has resulted in an increase in NDEs. So we have managed to lose the language and ritual around spiritual experiences of death—even as we have managed to increase the likelihood of hearing back from the near dead about their spiritual encounters.

One of the most intriguing cases for the medical staff involved occurred in Holland. A critical-care nurse was working the night shift when paramedics brought in a forty-four-year-old man in a coma. He had been found an hour earlier in a park, where a couple of Good Samaritans had tried to resuscitate him after a heart attack by pounding and massaging his chest. Once in the hospital, still without pulse, he was placed on artificial respiration and subjected to de-fibrillation procedures.

"When I want to intubate the patient," the nurse reported, "the patient turns out to have dentures in his mouth. Before intubating him, I remove the upper set of dentures and put it on the crash cart. Meanwhile, we continue extensive resuscitation." It took another hour and a half to stabilize the man's heartbeat and blood pressure. He remained comatose, and was transferred to the ICU where he hovered between life and death for over a week. When the patient regained consciousness, he was transferred to the cardiac ward. Said the nurse: "As soon as he sees me he says, 'Oh, yes, you, you know where my dentures are.' I'm flabbergasted. Then he tells me, 'Yes, you were there when they brought me into the hospital, and you took the dentures out of my mouth and put them on that cart; it had all these bottles on it, and there was a sliding drawer underneath, and you put my teeth there.' I was all the more amazed because I remembered this happening when the man was in a deep coma and undergoing resuscitation." The man explained that he had gone out of his body, and observed the medical team working on him. Apparently, he then offered a detailed account of what the room looked like, as well as the staff.

This has become known as the "dentures case," and it is often discussed now by those debating what near-death experiences mean. The other story often referred to is the "Reynolds case." Musician Pam Reynolds entered surgery in Atlanta, Georgia, in 1991 to remove a brain aneurysm that risked bursting and killing her at any time. Because of its location, the surgery carried great risks; forty or fifty years ago it would have been fatal. Now, however, she could be put into hypothermic cardiac arrest, with her core temperature dropped to sixty degrees, so that the aneurysm didn't hemorrhage while the surgery proceeded. Once she was cooled, all blood was drained from her brain, yet somehow, in the midst of her induced clinical brain death, Reynolds experienced herself as awake and out of her body. Not only did she not feel groggy—she felt sharper and more alert than she ever had in her life. (One analysis of the medical records of people reporting NDEs found that they described enhanced mental functioning significantly more often when they were actually physiologically close to death than when they were not.)

From an out-of-body position, Reynolds observed the unusual cranial saw that the neurosurgeons used to cut her skull, above and behind her eyes, which were, at any rate, taped shut. Reynolds later reported that the saw emitted a natural D tone. She noted the unexpected pattern they'd shaved into her blond hair, and heard the voice of a woman commenting that her femoral vessels were too small for the cardiopulmonary bypass shunt. She also reported that—after "returning" from an extraordinary encounter with light and deceased relatives—she saw her body "jump" twice. (She had gone into cardiac arrest postsurgery and was being defibrillated.) Reentering her body "felt like diving into a pool of ice water." (If people feel euphoria related to endorphin release, as is sometimes proposed as the explanation for NDE, the chemical effect shouldn't end abruptly and coincidentally at the exact moment when they have the psychological experience of returning to their body. Endorphins have been shown to circulate for hours.) She also heard the doctors playing "Hotel California," and she joked that the line "you can check out anytime you want but you can never leave" was "incredibly insensitive" given her distress at having to return to her prone form from the marvelous peace of the NDE.

Everything she reported that she saw and heard proved true. Reynolds's neurosurgeon, Robert Spetzler, would later say to the

BBC: "I don't think the observations she made were based on what she experienced as she went into the operating theater. They were just not available to her. For example, the drill and so on, those things were all covered up. They aren't visible; they were inside their packages. You really don't begin to open until the patient is completely asleep so that you maintain a sterile environment."

Keith Augustine, a young philosopher and avowed skeptic, reviewed the Reynolds case and argued that she could have seen things later and retroactively described them; she could have heard the music, the sawing, and the conversation through ordinary means, somehow rousing from the anesthetic while eluding the monitors that were tracking her brain waves. To this, the psychologist Janice Miner Holden responds: "Ms. Reynolds' brain was being monitored three different ways to ensure that she was deeply and consistently anesthetized. One of these ways was the monitoring of her most basic level of brain function by stimulating her hearing with ear speakers. Augustine did not report that the ear speakers molded into Reynolds' ears emitted clicks throughout her entire period of general anesthesia at a loudness of 90 to 100 decibels and a rate of 11 to 33 clicks per second. That volume has been described as [being] as loud as a lawn mower." It took that level of noise just to ensure that the patient was *deeply unconscious*. Imagine being profoundly sedated by general anesthetic with your ears sealed and *filled with lawn mower sounds that you were too unconscious to register*—and yet still managing to hear doctors mutter about an arterial vein. It's not completely impossible—after all, my musician husband points out to me, deaf people can play music by sensing vibration. But it's safe to say that the Eagles weren't playing live music in the operating theater that day, so the vibrations would have to have been coming from the radio and from the muttering surgeons.

The only alternative explanation is that she heard about her procedure in great detail after the fact, and decided, for some reason, to weave a false memory into an imagined encounter with the divine.

British psychologist Susan Blackmore has circulated a theory about what happens in NDEs, which she has called "the dying brain hypothesis." The perception of going out of your body, she has argued, "is the brain's way of dealing with a breakdown in the body image and model of reality" when near death. This may be due to pain, injury, or "it may be that the brain is no longer capable of building a good body image even if it had the information because it is ceasing to function properly." To construct a body image in this state, Blackmore said, we draw on memory and situate ourselves at a "bird's-eye view" of ourselves. "Memory can supply all the information about your body, what it looks like, how it feels and so on. It can also supply a good picture of the world. 'Where was I? Oh, yes. I was lying in the road after that car hit me.'"

Her critics complain that Blackmore's hypothesis is too speculative, that more evidence would need to be accumulated to support this idea that people remember themselves from the vantage point of the ceiling. They object, further, to Blackmore being so definitive about her idea: "At last we have a simple theory of the OBE," she has said. "The normal model of reality breaks down and the system tries to get back to normal by building a new model from memory and imagination. If this model is in a bird's-eye view, then an OBE takes place."

A lightly stated dismissal of any remaining mystery around the OBE appears to be premature. There is still no consensus as to what happens to people experiencing themselves as being out of their bodies, or even whether they are all experiencing the same thing.

Swiss neuroscientist Olaf Blanke, who has studied the sensed presence phenomenon, has also researched neurological patients, using brain scans, to probe the nature of OBEs. He suggests that the perception of being out of one's body may be the result of "paroxysmal cerebral dysfunction of the temporoparietal junction (TPJ)." That is to say, it may have to do with damaged neural wiring, involving an effect somewhat akin to seizures. Other scientists aren't persuaded. "The interpretation of the empirical findings of Blanke and his colleagues is controversial," writes the neuropsychiatrist Michael Kelly. The main objection, apparently, is that most people who have spontaneous OBEs haven't been shown to manifest this particular brain dysfunction. "The generalization from these few patients with identified neurological problems," Kelly and his colleagues at the University of Virginia caution, "to all persons experiencing OBE, most of whom have no known neurological problem, is purely conjectural."

The same note of caution is sounded about OBEs being understood as a symptom of epileptic seizures. The Austrian neurologist Ernst Rodin, who specialized in epilepsy before his retirement, has said that, "In spite of having seen hundreds of patients with temporal lobe seizures during three decades of professional life, I have never come across that symptomology (of NDEs) as part of a seizure."

What the arguments amount to is a continuing lack of clarity about what is really going on. "People are pushing the boundaries to extremes to try to put a label on this," Dr. Sam Parnia wrote in his 2013 book *Erasing Death*. He offers the example of a study done with student volunteers, outfitted with special goggles that showed them an image projected by a camera that had been stationed behind them. They were staring, in other words, at a picture of their own backs inside the goggles. The volunteers were left in the goggles for some length of time. "After a while they became so used to

that image that it felt like they were looking at themselves from behind. Then the researchers pretended to attack the camera with a
hammer. The people were startled because for an instant they felt
like someone was attacking them." From this, Parnia complains,
"The researchers concluded that they had reproduced an out-of-body
experience in the laboratory. Of course, this is not even close to what
someone who is critically ill, suffers a cardiac arrest, is resuscitated,
and then describes hearing conversations or seeing events in an out-
of-body experience would go through . . . I even wondered whether
[the researchers] had actually ever met and interviewed people with
out-of-body experiences."

Parnia has found in his own ongoing research in several collaborating hospitals in the United States and Europe that OBEs are being
reported by around 2 percent of the cardiac arrest survivors who
are medically rescusitated in hospital. In other words, of every hundred patients successfully brought back from flatline in the collaborating wards, two patients will have recall of exiting their bodies and
observing what is transpiring below. (More than that report NDEs, but
without the OBE feature.) It's important to note that this is different
from the overall prevalence rate of Americans who have had NDEs,
which is tallied over time, and includes people like David Bennett
who weren't rescusitated in a hospital. Who reports an NDE in
Parnia's research and who does not may have to do with brain trauma
in relation to the cardiac arrest itself. Cells in the hippocampus, the
area of the brain considered crucial for the formation of memory,
are particularly sensitive to damage. Both head injuries and oxygen
deprivation, known as anoxia, usually feature amnesia around the
incident. *"I don't remember falling." "The last thing I recall . . ."*

Parnia is wary of chalking up the near-death experiences generally to oxygen deprivation. In his one-year study of cardiac arrest

survivors at Southampton General Hospital in England, before he moved to the United States to work at Cornell, those who had NDEs showed slightly elevated levels of oxygen relative to the patients who reported no memory of NDE. "Our sample didn't seem to support the concept that NDEs were being caused by a lack of oxygen to the brain," he wrote. "From a medical point of view, lack of oxygen is a very common problem in a hospital. Most doctors working with emergencies come across it regularly, particularly in patients whose lungs or hearts aren't working very well—for example, in cases of severe asthma or heart failure. . . . When oxygen levels fall, patients become agitated and acutely confused. This 'acute confusional state,' as it is known medically, is very different from the near-death experience. During it, people develop 'clouding of consciousness' together with highly confused thought processes with little or no memory recall. . . . If the dying brain theory were correct, then I would expect that as the oxygen levels in patients' blood dropped, they would gradually develop the illusion of seeing a tunnel and/or a light. In practice, patients with low oxygen levels don't report seeing a light, a tunnel, or any of the typical features of an NDE."

In 2005, the University of Kentucky neurologist Kevin Nelson proposed that a near-death experience is something more akin to a waking dream. Specifically, people who had NDEs seemed more susceptible to a dream intrusion when their fight-or-flight system was aroused. People experiencing the terror of night paralysis have the impression of being awake and observant while actually in the dream state known as "rapid eye movement," or REM. This idea garnered lots of media attention, although experiencers interviewed for comment said, *Back off* my NDE. Nelson has publically described NDEs as "spiritual experiences," but he thinks they can be scientifically accounted for.

The radiologist Jeffrey Long and the psychologist Janice Miner Holden responded to Nelson that 40 percent of the NDEers in their survey had actually said "no" to questions about whether they ever experienced REM intrusion. How, then, could it be an underlying cause? Meanwhile, the Brown University researcher Willoughby Britton found—in an unrelated study—that, after their NDE, people entered REM sleep almost sixty minutes later, on average, than other sleepers. Once again, we have theories and counter-theories, as the evidence gets volleyed back and forth across the net.

Ultimately, we are brought back around to the meaning of the experience for those who have it, and what it tells us about what the dying may feel. It is as profound and as resonant as a hero's journey, as complex and instructive as myth. And it feels epically real. Why would the realm of the numinous be *more* real than our earthly reality? Is it because all the senses are sharpened, the emotions heightened, and the awareness of truth radically enlarged? We don't know. It just is.

"I make my living as a clinical psychiatrist dealing with people who have trouble dealing with what's real and what's not real, what's not delusional," says the psychiatrist Bruce Greyson. "And I've also spent thirty years dealing with people who've had near-death experiences, who have told me that this world may be real, but *that* world is more real. So, I've done a lot of thinking about 'what is reality?' We don't use, as our criterion, what's going on in our brains. People say, is it subjective or is it objective? But, in fact, that's not how we talk about reality. We all talk about emotions being 'real' or being feigned. Is that a real pain or is that a fake pain? That's not an objective thing. Pain, love, hatred, arrogance, these are subjective things, but we call them real or not real."

When Nancy Evans Bush, who had a frightening NDE, was asked

what question she found most irritating to be asked, she said, "Probably the one I dislike the most is 'Do you believe these NDEs? Are they really true? Do you really believe near-death experiences?' It's such an annoying little mosquito of a question because it indicates just such a lack of thought. They are experiences! You can't ask people, 'Is your experience true?' any more than you can ask someone with an abscessed tooth if their experience of pain is true. You're having the experience; of course it's true—as a genuine experience. Now, what does it mean? That is something different. Do I believe these experiences? Of course I believe them. Do I believe they are literally true? That is a different question with a far more complex answer."

The uniqueness and power of the spiritual experience is what we are left with. As Bruce Greyson said in 2008 at the UN-sponsored meeting on the mind-body problem in New York, where he mused about the nature of reality, "We have no blood levels for enlightenment, but we can study its aftereffects." It is consistently found, for instance, that people who have NDEs are measurably changed by it, just as we are finding that people who encounter the presences of their deceased loves ones are measurably consoled.

NDEers become less dogmatically religious—if they were religious to begin with—and yet more actively spiritual. The percentage of Australians in Cherie Sutherland's study who claimed to have no religious affiliation prior to their NDE was 46 percent; that number jumped to 84 percent afterward. In other words, the NDE convinced the majority of people who had it that the doctrine they had been accustomed to—whatever it was—was off base. Pim van Lommel's Dutch research showed that church attendance declined by 42 percent in his NDE group, while spirituality nearly doubled. By contrast, his control group of cardiac arrest survivors who did not have an NDE showed a 42 percent decrease in interest in spirituality after eight years. (There may be no atheists in foxholes, as the saying goes, but many may revert to that philosophy once out of danger—unless they've encountered the numinous. *And then they don't.*)

So what does an active spirituality look like, then? Compassion and empathy for others increased by 73 percent at the eight-year mark versus 50 percent for the control group in Lommel's study. Interest in material status decreased by 50 percent; whereas that same interest—the pursuit of money, the quest for accomplishment—*increased* by 33 percent in controls. What is meant by spirituality here, I assume, is the cultivation of love, the knowledge that you have a purpose, and the acceptance that ego is nothing if not entirely irrelevant.

In May 2013, I caught up with David Bennett in Syracuse, where he had arranged for the neurosurgeon and bestselling author of *Proof of Heaven*, Eben Alexander, to give a talk about his own NDE. In person, Bennett—with his white-gray hair in a military cut and frameless glasses—had a pugnacious look that put me faintly in mind of the late actor James Gandolfini. A working-class New York guy, yet with a slightly haunted look about him. It takes someone who has had a deep NDE an average of twelve years to integrate the "radical shift in reality" into their lives, according to the psychologist Yolaine Stout, who has studied the impact of the experience. That had certainly been the case for Bennett. It wasn't until he was invited by friends to meditate in the Arizona desert in 1994 that he faced what had happened to him. In fact, he relived the entire NDE in the midst of meditation and was left sobbing and shaking on the ledge of a mesa. "It was like being hit by a two-by-four," he told me, "and I was like, okay, *okay*, I get it. Now I'm going to have to start living my life knowing what I know." When an acquaintance came across him weeping on the mesa, he told his story for the first time. Eventually he began running a support group in upstate New York.

In Syracuse, Bennett was wearing a blazer, without a tie, and seemed a bit overwhelmed by the number of people in attendance. Ordinarily, his group gets one or two dozen people out to their events, but Alexander's book had been on the bestseller lists for months. Hundreds of folks had filled a local Holiday Inn ballroom.

Like most people I've met who have had a spiritually transformative experience, Alexander had an air of calm about him when he took the podium. Perhaps when you've been told that you're loved by God and will always be loved, you lose that edge of performance anxiety that tends to plague the rest of us.

Not that Alexander is unself-conscious. He sports a signature bow tie, and comports himself with the assuredness you'd expect to see in a surgeon who has been financially and intellectually powerful for years. This isn't Francis of Assisi, discovering God and throwing away all his worldly possessions. Instead, and not insignificantly, Alexander threw away his worldview, or—at any rate—his prestige as a medical rationalist. (As one doctor said after his NDE, "walking around saying that you'd seen and heard things that weren't there wouldn't bode well with the Department of Health. Telling my colleagues was low on my to-do list.")

Alexander was a busy neurosurgeon in Virginia when, in the fall of 2008, he came down with an extremely abrupt attack of a rare bacterial meningitis, so virulent that he lapsed into a seven-day coma and had a roughly 3 percent chance of surviving, much less emerging without brain damage. But he did survive entirely intact, and, like St. Paul on the road to Damascus, he was thunderstruck by the spiritual revelation of an NDE. His experience began with what he chose to call the Earthworm's-Eye View.

"Time flow in that realm is very different, so I don't know how long I was in a murky underground place," he said, "like being in dirty Jell-O, listening to a pounding, smashing sound. It sounds fore-boding, but since I remembered nothing of my life, I accepted it." This description is reminiscent of the Egyptian Book of the Dead's description of the afterlife, which begins with a plunge into the muddy and disorienting realm of the Duat, before ascending to the higher realms. Actually a papyrus tucked in with a mummy, the "book" was to act as a travel guide or road map for the afterlife.

Alexander's NDE was unusual insofar as he cycled repeatedly be-tween three different realms—the murky place, a beautiful valley, and some sort of higher God realm. Maybe he was actually drifting between being in his diseased body and out, the way dying people sometimes do. Of the valley, he said with a slight wobble in his voice that has become familiar to me in people giving such testimonials, "It was *lovely*. It's very hard to put words on this." And of the more in-finite God realm, and the experience in general, he said, "I cannot tell you how *totally* reassuring that unconditional love is. It was absolutely pure and perfect . . . even when I was just researching this as a neuro-scientific experience I remained haunted by the power of that love."

Reading for the first time in his career the prevailing theories of what goes on in an NDE, Alexander was "astonished by how flimsy they were." The question is whether science can ever devise the right answer. "That realm is beyond our understanding," he said. "'God' is a teeny little human word. There's no way that creation can ever fully understand the creator." He also wanted to be clear that he didn't think of this realm as heaven in a Christian sense. Although he had a Christian upbringing, his NDE did nothing to convince him that Jesus Christ was the only prophet or savior. "In the four and a half years since my NDE, I have been trying to come up with a worldview that makes sense. We need to take down the artificial boundaries between religions, and between religion and science."

The essential quandary of our culture is how we approach a larger spiritual reality that is implied by deathbed experiences, near-death experiences, and shared death experiences when we're working merely with hints, glimpses, and guesses. For many, it's easier not to think about it. Life is complicated enough without all the uncertain complexity of what might ensue. We either want a quick reassurance that it's heavenly, or a declaration that it's swift execution—lights out, brain gone. It's just simpler that way. I was driving with my mother about ten years ago, and for some reason we got onto the subject of the afterlife. Confidently, Mum quoted an ancient Roman thinker: "Where death is, I am not; where I am, death is not." It sounded good.

That was before Katharine and Dad, and the unraveling of everything we knew. Now, after interviewing people like Yvonne Kason and David Bennett, and pushing my way through the thorny thickets of scientific debate and confronting my own prejudices and anxieties and basically doing all the fraught reckoning that I'd always avoided because *it's too complicated*, I am fonder of this quotation, commonly attributed to the Indian Nobel Laureate for Literature in 1913, Rabindranath Tagore:

"Death is not extinguishing the light; it is only putting out the lamp because the dawn has come."

There is nothing that we know scientifically one century later that says that he was wrong.

CHAPTER 7

Walking the Enchanted Boundary: Living in the Aftermath

*The more that critical reason dominates, the more im-
poverished life becomes; but the more of the unconscious
and the more of myth we are capable of making conscious, the more of life we integrate. Overvalued reason
has this in common with political absolutism: under its
dominion the individual is pauperized.*

—CARL JUNG

It is nearing Halloween in Rhinebeck, New York, about an hour
and a half's drive north of New York City. It's a place of perfect
American autumn, like something out of a Martha Stewart picto-
rial: white picket fences, neat sidewalks featuring artfully decorated
pumpkins, the air spiced by blazing oak and maple trees. Well-
groomed horses graze on hobby farm grass. At Halloween, the
dead are welcome here in tidy costume, or as expensively spooky
lawn ornaments, but this year they have been summoned in ear-
nest to the Omega Institute, a retreat complex in the forest near
Rhinebeck that holds weekend workshops on the art of meditation,

Reiki, and sound healing. This weekend is all about "soul survival," and it's a sell-out event.

In the main hall, surrounded by piney woods and organic gardens in late harvest, Raymond Moody sits in a winged armchair on a wooden dais backed by red drapes, one leg crossed over the other, lightly bouncing a foot clad in a black-and-white sneaker with neon green laces. The footwear is somewhat incongruous for this aging southern gentleman. Is he aiming for ultimate comfort, like the three hundred people who have strolled from their Spartan cabins in this old converted summer camp to hear him, many of them shell-shocked by recent deaths? There is a young woman who lost her fiancé two weeks earlier in a motorcycle accident in Brooklyn and felt driven to despair that she hadn't somehow known he had died, and had carried on for three more hours working her pharmacy job. There is a New Englander, neatly attired in a cashmere sweater set, who saw her father rise up in his bed upon death, as if his body were animated by a mysterious force. A Russian who had devoured Moody's first book when he came across an illicit copy in the officially atheist USSR is in attendance; a man of evident wealth who had a near-death experience during a skiing accident in Aspen and now felt lost, with one foot "in both worlds." Turned away by the scientific, medical, and media establishments, as well as by the many churches that frown upon spontaneous spiritual encounters with the dead, they wind up congregating at events like this.

"What the universe is interested in is how you have learned to love," Moody assures the audience, who sit on a combination of cushions and chairs with notebooks and cups of coffee. With his southern lilt and high-pitched voice, he sounds a bit like the novelist Truman Capote, diminutive, an outsider, smart. After describing what he's heard from thousands of people over thirty years of probing the NDE, he's become a sort of sage. "Whatever people were chasing before—fame, power, money—they come back knowing that what matters is love," he explains. He talks about the struggle to live with what you know, how elusive the path can be. "They know what the goal is," but not necessarily how to achieve it. "It's very hard to get through the average day without wanting to choke at least one person." The audience chuckles ruefully.

Everyone has difficulty achieving the goal of being perfectly loving, but one thing NDEers do achieve is a certain fearlessness, Moody says. He has seen NDEers die years later, retaining their "absolute lack of fear of death." He muses over other characteristics he's seen, citing the unusual cases of "swan song," in which people sing or recite poetry at the time of their dying. "Her face was going gray," he recalls of one patient he was trying to resuscitate in medical school, "my resuscitation didn't work, and yet she was reciting poetry." (This sounds to me like terminal lucidity, which in turn mirrors the enhanced perceptions and senses experienced during mystical encounters and NDEs. As the brain dies, the senses become unbounded.)

Although he is a psychiatrist, Moody is now acting, not unlike Eben Alexander, as a peculiarly modern kind of priest. Where Alexander is a charismatic preacher, Moody is more a kindly village friar, providing the counsel and reassurance that people who have encountered the numinous are hungry for, offering the research wisdom that leaves them feeling less stigmatized, less credulous, less psycho.

"In my experience as a psychiatrist, everyone is crazy," Moody says soothingly. "When people who've had NDEs ask me if they're normal, I say, 'What's normal?' Normal is somebody you don't know very well. Some of the people who have NDEs are the least neurotic people I know." If there is gratitude in laughter, this is what I now hear. Everyone, myself included, feeling grateful to be understood, relieved to be with the tribe.

Moody asks the audience to raise their hands if they have had a shared death experience. Twenty-three hands waver in the air, about a tenth of the people there. "Why would the people at the bedside have the same experiences? It doesn't make sense," Moody says. One woman in the audience describes a memory of her mother lying in grave condition in the ICU. She curled up on the hospital bed with her all night long, and at some point began sharing a panoramic review of her mother's life. She thought she was hallucinating, but when her mother emerged from a coma some days later, she described the same experience, the same witnessed details.

"We're into things here that are very difficult to explain away," Moody declared.

It is crazy hard to navigate the edges of what's true and not true, what's possible and what's not possible, as those edges blur and shimmer along an enchanted boundary between skepticism and belief. A medium at Omega, an affable and witty Bostonian named John Holland, told us stories of his awkward childhood where he saw things he shouldn't have been able to see, that weren't supposed to be possible. Having been belittled and shunned by other kids, he ignored his capacity for years, because it brought him nothing but derision. Working in L.A. as a bartender, he would periodically vent his bitterness at how the world had made fun of him by abruptly "reading" someone sitting at his bar—knocking them off their stools with unbidden secrets, like "Your aunt just died." It was a kind of psychic mischief-making. A friend persuaded him to cut it out, grow up, and train his gift, so he attended a school for mediumship in England.

At Omega, he prepared for a group reading. First, he bid us go for a walk in the woods, ask our loved ones to come if they wanted, to speak if they needed to; he cautioned us not to approach our dead with need, but rather with love. Out in the autumn sunlight, I walked and asked my sister to come if she wanted, if she wished me to write this book or should I quit it, just get on with my life, "break the bonds," as grief therapists used to implore. Should I stay or should I go? I had no confidence in what I was doing as real. It felt pretend. The Enlightenment Scot in me was fully armed. And yet, at the same time, this was about my own dear sister, one of my best friends. What if she *was* there, somewhere, and I was too proud to listen?

Back in the group, Holland explained that he would be drawn to various places in the room. At some point in the reading he came over to my area, and said, "For someone here, in this section, I'm seeing an old-fashioned radio." Immediately, behind me, a middle-aged woman with shaggy dark hair cried out that she volunteered for a radio station. He began offering her images which seemed, impossibly, to relate to me. These are the ones I remember:

"Okay," Holland said. "The left side of me is sagging. I'm seeing a breast cancer, left side, a mastectomy."

"I had a breast cancer scare!" the woman cried.

"This wasn't a scare," Holland said, "this was fatal."

The room was silent. I could not raise my hand.

"Okay, I'm getting Paris," Holland said, and although Katharine was born in Paris and I'd just been there for this book, I stayed quiet. The woman behind me called, "I'm thinking of going to France next summer, to the south!"

"It's not southern France," argued Holland. "It's Paris." He paused, seeming to listen to a disembodied voice. "I'm seeing a thyroid condition."

Just before Katharine died, I was diagnosed with hypothyroidism.

———

It was one of the last things, in essence, she knew about me. I remained silent.

"Okay," Holland said, clearly frustrated, but giving it one last go. "When is the book coming out?"

I sat there with my head thrumming, suddenly remembering my last conversation with Katharine before she went to hospice. We had talked about a book we both read by the fantasy novelist Connie Willis, called *Passage*. The main characters are scientists investigating near-death experiences by inducing the state in one another and monitoring the results. One of them crosses the threshold, and dies. She finds herself on the deck of the *Titanic*, a metaphoric construction for where she actually is, like Ellie Arroway's Florida beach in *Contact*, and she tries desperately to signal back to her fellow research scientists through the ship radio.

The old-fashioned radio.

Adrenaline rushed through me. I waited in line to talk to Holland, who was now signing books. "I just wanted to check with you," I said, when I finally came to the front of the line. "You had a message from someone who died of breast cancer after a left-breast mastectomy, associated with Paris, and a radio, and thyroid. My sister was born in Paris and died of that cancer and I have that thyroid condition and am writing a book about her, and we promised to communicate through an old wireless radio. Could the message have been for me?"

He stared at me, his bushy eyebrow arched in gobsmacked amusement: "Why didn't you *raise your hand*?"

"I'm shy," I joked. But the truth was that I hadn't been able to. And now I was left with this insane ambiguity that was making my blood sing. The book-signing queue pressed in from behind so I nodded my thanks and went forlornly outside.

"Come on," said my friend who'd come down with me for the weekend, "let's go get a drink." But I couldn't just drop this and shrug. It would be like thinking your fiercely adored sister has just banged on your front door, and you know she's been lost and have been longing to hear from her and help her, but on the other hand it was probably the wind banging the door so just go out of the back of the house and attend that cocktail party you'd agreed to go to. *Like that.* If you take this thing a little bit seriously, you have to take it all the way seriously or it messes with your head.

Carl Jung wrote about this dilemma in his book *Memories, Dreams, Reflections*:

"One night I lay awake thinking of the sudden death of a friend whose funeral had taken place the day before. I was deeply concerned. Suddenly I felt that he was in the room. It seemed to me that he stood at the foot of my bed and was asking me to go with him. I did not have the feeling of an apparition; rather, it was an inner visual image of him, which I explained to myself as a fantasy. But in all honesty I had to ask myself, 'Do I have any proof that this is a fantasy? Suppose it is not a fantasy, suppose my friend is really here and I decided it was only a fantasy—would that not be abominable of me?' Yet I had equally little proof that he stood before as an apparition. Then I said to myself, 'Proof is neither here or there! Instead of explaining him away as fantasy, I might just as well give him the benefit of the doubt and for experiment's sake credit him with reality.' The moment I had that thought, he went to the door and beckoned me to follow him. . . . That was something I hadn't bargained for. I had to repeat my argument to myself once more. Only then did I follow him."

Jung envisioned himself following his friend, the newly deceased physicist Wolfgang Pauli, to his nearby house and into his study, where Pauli climbed onto a stool and pointed to "the second of five books with red bindings which stood on the second shelf from the top." Having made this gesture, Pauli and his study vanished. The next day, Jung asked Pauli's widow if he could look in his friend's library. "Sure enough, there was a stool standing under the bookcase I had seen in my vision, and even before I came closer I could see the five books with red bindings. I stepped up on the stool so as to be able to read the titles. They were translations of the novels of Emile Zola. The title of the second volume read: 'The Legacy of the Dead.'"

Traveling back to Toronto from Omega, the train I was riding collided with a car.

It was 10:30 a.m. on a Monday. I was transcribing notes into my laptop about Moody's comments on shared death experiences. Suddenly, I registered a muffled bang or thump. I was in the back of the train, so the sound was remote. Had we hit a deer? We came screeching to a halt, and I looked up from my computer and gazed out the window at a harvested cornfield in silvery light. This would, as it turned out, be my view for the next five hours.

Moments after the stop, I smelled smoke and burning rubber, and wondered if I was actually going to have to act on the emergency instructions I'd received from the steward that morning, to break the glass at my designated exit window with a ball-peen hammer. Was the train on fire? Wow. An adventure. But no. In fact, unbeknownst to us, a black Ford Escape was in full meltdown behind us on the tracks, thoroughly accordioned and spun. A twelve-year-old boy lay dead in the cornfield while his mother struggled for breath through fractured ribs.

The train-crossing gate hadn't worked properly, hadn't lowered; she had noticed our approach and slammed on the brakes a fraction of a second too late. These details I learned a couple of hours into our stop by searching the local news on my laptop. All that we passengers officially knew was that we'd come to a stop and could smell smoke. A train official crackled onto the loudspeakers, and warned us to expect a delay.

Cell phones got flipped open and meetings in Toronto were postponed or put on notice. The woman across the aisle from me phoned her trip-cancellation insurance provider to clarify her options, in case she missed the afternoon charter flight to her all-inclusive hotel in Punta Cana.

People sighed, joked about the train service, rolled their eyes.

A steward came through with a trolley cart of coffee and snacks, looking haunted and acting irrationally. To some, he sold sandwiches, and to others he said: "We're just doing juice here, just drinks. We'll deal with food later."

Had he been up at the front, yakking with the train conductor, when together they saw the car approaching, pressed a useless, urgent bell, and then watched a child fly through a windshield?

Likely yes, for the French-Canadian steward sought out anyone with a passable knowledge of French in my car and proceeded to rant to them about what he'd seen on the job, the suicides, the accidents, the reckless attempts to rush the crossing, *"C'est fou,"* he declared. Crazy.

No, what was crazy was that a child who had been alive when we boarded this train was now lying mangled and lifeless fifty yards behind us and we couldn't find out about it except through our smart phones, and then, when we did, we couldn't figure out a collective moment of silence, or a brief ecumenical prayer for the family or his soul. There we sat for five hours, first waiting for the paramedics and police, and then for the coroner, and in all those hours no one proposed to get off the train to pay our respects to this death, or ac-knowledge that it had happened in any way, other than referring to our interrupted schedules.

When we got moving again, the man across the aisle said, "You know, people have been surprisingly patient about this delay," and I said: "Well, maybe that's because they realize that the delay is trivial compared to being killed by the train."

And he gave me an apologetic smile. "Oh, that's probably right."

We finally arrived in Toronto at 5:30 p.m., and I had to go straight up to U of T for the class I was teaching. I asked one of my students, who is an observant Muslim from Pakistan, what would have hap-pened in similar circumstances in Lahore, when he was still living there. "What if a person was killed right in front of you, would you ignore the death, sigh, and rearrange your flight to an all-inclusive in Punta Cana?"

"No," he said, laughing in surprise, "we would rush to tell the imams, and every mosque nearby would begin ringing bells."

If we can't even face the basic fact of death in secular North American culture, how can we be expected to understand and re-spect the radical profundity of it, and thus begin to sort out the "Psy-chic Fayre," as the novelist Hilary Mantel described our indulgence in "fun-size beliefs," from the true mystical experience, the NDE, the sensed presences? We deny, or we radically accept, but we have no frame, no wisdom. *Where is the wisdom we have lost in knowledge? Where is the knowledge we have lost in information?*

A week of rain and fog ensued, during which I encountered some unsettling teachings from *The Tibetan Book of Living and Dying*, by Sogyal Rinpoche. In the stone monasteries of the Hima-layan foothills, it is a common belief that when you die, you enter what are known as the bardo states. Buddhist monks prepare studiously for this experience of entering the bardos, because the states—or realms—of the bardo can be trippy in the extreme. They might be ecstatic, but also hellish, and if you don't keep your cool as you encounter demons, wrathful deities, and monstrous wolves, you will race for the nearest available exit—which could result in rebirth as a stick insect or an aloe vera plant. I simplify, but, psychologically, you have to run a bit of a gauntlet before you get to any kind of heaven.

I took the dog for a walk in the drizzling evening, the air redolent of rain-damp oak leaves, and wondered uncomfortably what this meant for Katharine and Dad. Near-death experience accounts are profoundly comforting: you die, and find yourself in the best place ever. But Tibetans insist that that's too simple and too easy. Instead, a much more complicated and potentially frightening experience ensues that is difficult to navigate without studious advance practice. Yes, they say, your emotions are heightened, but not only the good ones. And if you give in to your emotions, or at least to the negative ones, you will be swiftly ushered into your new stick insect costume: off you go to someone's back garden in Coventry.

I contemplate this formulation as I walk along Bellwoods Avenue in downtown Toronto. My middle-aged sheltie ambles a pace behind me, immersed in a world of immediate, present scents—urine, grass, a pizza crust—every step a marvel of information. A young man strides toward and past me, his focus glossed over by whatever is on his iPod. He is present, but not present, taking advantage of what he figures is a safe external environment to space out on his musical choice. In the bardo realms, you don't space out. You *are* spaced out, literally. And you have to concentrate.

How do you get there from here? If the Tibetans are right, then only they know how to brace for the experience of death, which means that the vast majority of the world's population will be blown immediately into the portal that rebirths us as beetles.

How does that make sense?

Could some monk not have passed along the word?

Nancy Evans Bush, who spent years contemplating her frightening near-death experience, writes, "We have to recognize that the universe is made up of darkness as well as light, so we'd better pay some attention to the implications of that." It's not just heaven or lights-out. Every mystic and saint, every hero has had a more harrowing journey. On some level we know that. "The deepest enigma for human beings [remains] learning to live with what we believe. That's the hard part," she wrote.

It is hard—or all the harder—when we abandon our wisdom traditions.

One day, four years after Dad and Katharine died, I took the children to our lake cottage. This is bedrock home to me and my clan, the place we come from and go to, where we draw our strength. I was awakened the first morning by someone calling my name, or at any rate: "Pat." Since my son, Geoffrey, often calls me that now, teasingly, I assumed he was trying to get me out of bed. I roused myself and went out onto the porch, where he was deeply immersed in a computerized game of chess.

He hadn't said a thing.

Oh.

My daughter, Clara, woke up a few hours later. She'd been struggling to sleep for the previous couple of nights due to a respiratory infection. She discovered an extra blanket laid atop her.

"Did you put that on me?" she asked.

"No," I said.

Oh.

In the afternoon, when Clara was having a nap, she felt someone gently grasp her foot. A sensed presence, although she didn't know that term.

Toward midnight, I sat at the table on the screened-in porch listening to the whine and whir of insects banging against mesh. The full moon rose through the branches of the pine tree outside, ascending with cool beauty through shadowy clouds. Choral music was coming from the speaker and I thought of my sister, who had loved Clara, and would have snuggled her up with blankets. Clara had sung at Katharine's memorial service. Proud of her exquisite musicality, I prompted her to repeat the song later at a dinner party and she fixed me for a fool: That was sacred. Unrepeatable. That was for *Aunt Katharine.* She left the room.

In the dark, familiar quiet of my cottage I abandoned my defensive musings and my journalist's distance and thought about my sister openheartedly. Had she not come to me? When I lay in bed some weeks after she'd died and asked in messed-up anguish for a sign that she was all right, I found a single vivid pink bloom—not a bud, but a ridiculous, time-lapse bloom—on a long-dead plant in the hallway. When I asked her to come to the Omega Institute, the medium relayed details that were neatly matched to me, even though I wouldn't claim them. And now, I heard knocks and whispers.

We reach and reach, trying to stay connected across space and time. I was reflecting on this genuine fidelity of Katharine sending her radio messages to me since I'd lost her—when the Sharpie on the table near my laptop suddenly rolled.

Almost immediately, I began to doubt the truth of what I saw. Or, no: I *did* see the Sharpie roll, there is no question, because it startled me to see it. I tried breathing out a puff of air, to test if it would roll again. There was no breeze, but perhaps it had to do with the way I was breathing? That's what I figured. But no. Nothing, since. It didn't move again. Any more than the phone book or the bowl of fruit or the other pen nearby had moved. Only the insects continued to bang against the porch screen.

The Sharpie firmly and decidedly moved across the table about three inches, as I thought about my sister's persistence in attempting to reassure me that she was alright.

I am so primed by my culture to disbelieve in the significance of this signaling pen that I picture my sister throwing up her hands— "How much more obvious can I get?"

Never before has a pen moved across a table in front of me; never have I hallucinated. As I had begun reflecting on my sudden sense of genuine gratitude for her efforts over the past four years, I had started *believing* in her. Whereupon the pen moved. What had the medium John Holland said? You need to approach the dead with love, not with need.

Yet if I am to accept that my sister communicated with me, then I have to start contemplating where she is, what she has been up to. Is she in the Bardo realms? Totally foreign! I have no cultural understanding of that. Is she in a new body? Is she a baby in Nicaragua? Has she become a formless drop of light, and if so how will I find her? When the philosopher Ken Wilber lost his new wife to cancer, she kept telling him repeatedly as she was dying, "Promise me that you'll find me." How do we *find* one another again?

All these questions, so painful, lead us to refusing to answer, which is so much easier. A couple of years ago, I walked along the cliff edge of the Adriatic Sea, along a trail that the great German poet Rainer Maria Rilke once paced a hundred years ago. In a letter penned above these limestone rocks, he wrote: "This is in the end the only kind of courage that is required of us: the courage to face the strangest, most unusual, most inexplicable experiences that can meet us. The fact that people have in this sense been cowardly has done infinite harm to life; the experiences that are called 'apparitions,' the whole so-called 'spirit world,' death, all these Things that are so closely related to us, have through our daily defensiveness been so entirely pushed out of life that the senses with which we might have been able to grasp them have atrophied."

Yes, they have. Of course they have.

Sitting in the boat at the lake house in the autumn of last year, bailing leafy water out of the stern with a plastic jug: a rhythmic scoop and pour, scoop and pour, cool water rushing in, sloshing over, trickling, dripping, creating all this watery acoustic sound in the otherwise uncommon silence. Sun casting a pure, high light across the lake, air scented by brackish water. Silence and sunlight and water.

There is sometimes a juxtaposition of those elements that approaches the sacred, and I can only imagine what the numinous light would be like. As Teresa of Avila said, this water, and this light, is but a muddy stream in comparison. If Katharine were trying to communicate with me, it would only be to say, "You have no idea—there are no words!" As palliative psychologist Kathleen Dowling Singh has noted, "The dying become radiant, and speak of 'walking through a room lit by a lantern,' or of their body 'filling with sunlight.' They share these experiences with a quiet, thankful awe. There is something about the quality of their demeanor and expression at these times that has a feeling of purity, like a world wiped clean by new snow or as seen through the eyes of an infant. Ego is not in these expressions . . . It would appear that we move into a sacred realm of Being that we recognize with faith, confidence and gratitude to be our own. One woman kept returning from this depth of being to let those of us around her know: 'I cannot tell you how beautiful this is.'"

I meet up with my sister Anne and her husband Mark, and we spend the afternoon tucking the cabin up for its winter slumber. Much of what we do means confounding the squirrels, who appear to have spent most of their autumn collecting acorns and carefully hiding them in our bed sheets and between the folds of swimming towels. Each time we strip a bed, acorns tumble out of hiding spots and bounce across the floor. It's like we're on an inadvertent Easter Egg hunt in each room. Anne and I keep laughing. We pry apart the docks by bolt and screw, and use the boat to tow them around to the lee of the boathouse bay, where Mark ropes them to cedar bushes and we hope for the best.

The next day, as I shutter the windows to the cottage for the season, I wonder what will have happened when they are next thrown open to soft spring light. What will have transpired in my life, in ours, in the history of the world? Who else will have died?

But the grace I see, now, comes from the comfort I draw from my tribe, my sisters and brother, cousins, aunts, and uncles, extended family, and friends. Love is the coin of the realm, said Eben Alexander, and he's right. We can't know what comes next, what beauty or terror or hero's journey, but we can draw our intuitive wisdom about how to live from what we hear and see. The family has drawn ever closer, unself-consciously so, through the astonishments of death, and the new confidings about our respective experiences with the mystic.

When the dying leave us, it's like a footprint in the sand that needs to be filled in. Where the water rushes in, where love rushes in.

Acknowledgments

This book has been fiendishly difficult to write, for a number of reasons, not least of which is the wall of skepticism I've had to scale even in my own neighborhood, among my own friends. I am profoundly grateful to those who have encouraged me.

I would like to thank, above all, my mother, Landon, who permitted me to share the story of my father's and sister's deaths without attempting to control the narrative; who introduced me to various key sources, including the German theologian Rudolf Otto; who read drafts and offered incisive feedback; who supported me in every imaginable way with open-minded curiosity and heart. For a woman who has achieved international respect as an architect and advocate for the United Nations' Declaration of the Rights of the Child, she has practiced what she preached in her own family. I have never felt anything less than completely engaged respect from her as I've fumbled my way along, seeking meaning.

Huge, respectful thanks to the women who fought for this project to come to fruition, walking around kicking shins so that contracts could come into being: my amazingly smart and witty agent, Sarah Lazin; my editors—Anne Collins at Random House Canada and Leslie Meredith at Atria—and my longtime publisher and friend Louise Dennys. I am honored by you all for giving me voice. Also, I hate you all but respect you all for making me rewrite this book so many times. It has been the great pleasure of my career to work with fiercely intelligent, utterly sane, and uncompromising women.

None of this writing would have been possible without the foundational research of some key people who have risked their reputations to keep our inquiry into human spiritual experience going in a time of heightened materialist belief. In particular I am indebted to my fellow journalists John Geiger and Maria Coffey; to the psychologists Erlendur Haraldsson and Karlis Osis; to the neuropsychiatrist Peter Fenwick; to the many excellent scholars at the University of Virginia's Division of Perceptual Studies, including the late Ian Stevenson, Bruce Greyson, Emily Williams Kelly, and Edward Kelly; to the doctors Penny Sartori, Sam Parnia, Michael Sabom, and Pim van Lommel; and to members of the International Association for Near-Death Studies, in particular Jan Holden and Diane Corcoran.

Many people contributed their experiences to this book, and I am grateful to those who allowed me to name them as well as those who preferred to remain anonymous.

For editorial feedback and source suggestions along the way, my thanks to Ambrose Pottie, Marni Jackson, Judy Finlay, Teresa Toten, Dr. Mary Vachon, Sheila Whyte, Monique Séguin, Leah Cherniak, Manuela Jessel, Mary Mackenzie, Joanne Thomas Yaccato, Lewis Humphreys, Doug Campbell, Anne Pearson, Mark Vorobej, and Sarah Jordison. I am forgetting a million people, I know it.

Jeff Warren has been my sounding board and musing pal from the outset and, I hope, for many years to come.

My thanks to the editorial team at Atria for their close collaboration with me in production, particularly Donna Loffredo, Jennifer Weidman, and Stephanie Evans.

I am indebted to the Canada Council for the Arts for supporting this project.

Finally, I apologize to my husband and children for festooning the house with books about death for several years. They eventually got so used to it that Clara, at sixteen, suggested this book's title: *YOLO, or DO you?*

Notes

AN UNEXPECTED VISION

2 *Katharine's fate had become the family's "extreme reality"* . . . Virginia Woolf, "Sketch of the Past," in *Moments of Being: Autobiographical Writings*, ed. Jeanne Schulkind.

5 *"Beauty is only the first touch of terror"* . . . Rilke, *Duino Elegies*.

7 *"One has never seen the world well,"* wrote the metaphysicist Gaston Bachelard . . . Bachelard, *The Poetics of Reverie*.

11 *Swiss geologist Albert Heim, who fell off a mountain and wrote, in 1892* . . . A. Heim, "Remarks on Fatal Falls," trans. R. Noyes and R. Kletti, *Yearbook of the Swiss Alpine Club* 27 (1892): 327–37; "The Experience of Dying from Falls," *Omega* 3 (1972): 45–52.

11 *A study conducted by Harvard researchers found that 63 percent of doctors caring for terminally ill patients wildly overestimated how much time their patients had left* . . . See discussion of this prognosis error in Marcus Alexander and Nicholas A. Christakis, "Bias and Asymmetric Loss in Expert Forecasts: A Study of Physician Prognostic Behavior with Respect to Patient Survival," *Journal of Health Economics* 27 (2008): 1095–1108; see also Atul Gawande, "Letting Go: What Should Medicine Do When It Can't Save Your Life?" *New Yorker* (August 2, 2010).

15 *Palliative care physician Michael Barbato designed a questionnaire for family members* . . . M. Barbato, "Parapsychological Phenomena Near the Time of Death," *Journal of Palliative Care* 15 (Summer 1999): 30–37. "Our data are consistent with other studies," Barbato notes, "that show no correlations between religion, spiritual beliefs, or cultural background and the occurrence of parapsycholgical experiences."

17 *In 1979, a survey of more than a thousand college professors in the United*

States . . . M. W. Wagner and M. Monnet, "Attitudes of College Professors toward Extrasensory Perception," *Zetetic Scholar* 5 (1979): 7–17.

17 *In 1999, the psychologist Charles Tart put up a website called the Archives of Scientists' Transcendent Human Experiences . . .* Charles T. Tart, ed., *The Archives of Scientists' Transcendent Experiences*, Institute for the Scientific Study of Consciousness, http://www.issc-taste.org/index.shtml.

17 *A recent study in the British Journal of Psychology showed that there is no difference in critical thinking skills . . .* C. Roe, "Critical Thinking and Belief in the Paranormal: A Reevaluation," *British Journal of Psychology* 90 (1999): 85–98; see also Dean Radin, *Entangled Minds.* "Other studies confirm this lack of difference," Radin reports. "A 1997 study by Uwe Wolfradt in the journal *Personality and Individual Differences* found no correlation between dissociative behavior and belief in psi. On the other hand there is a strong correlation between absorption—the ability to concentrate or focus—and experience of psi . . . in spite of evidence to the contrary, some skeptics continue to assert that belief in the paranormal is best explained by ignorance or mental deficiency," p. 40.

17 *Better-educated people more likely to accept "psi". . .* The Harris Poll #11, February 26, 2003, Harris Interactive, http://www.harrisinteractive.com/vault/Harris-Interactive-Poll-Research-The-Religious-and-Other-Beliefs-of-Americans-2003-2003-02.pdf. In other words, it is pretty widely accepted that there is something going on in the universe that draws on another sense than the ones we've officially identified so far.

17 *The retired Princeton physicist Freeman Dyson wrote in . . .* Freeman Dyson, "One in a Million," *The New York Review of Books* 51, no. 5 (March 25, 2004). *"If one believes, as I do, that extrasensory perception exists but is scientifically untestable" . . .* Freeman Dyson, foreword to *Extraordinary Knowing: Science, Skepticism, and the Inexplicable Powers of the Human Mind*, by Elizabeth Lloyd Mayer (New York: Bantam, 2007).

18 *Dyson received flak for his assertion, but like many of us he'd witnessed inexplicable phenomena within the confines of his own extended family . . .* Ibid.

18 *As Dyson said of his cousin and grandmother: "Neither of them was a fool." . . .* Ibid.

19 *[University of California psychiatrist Elizabeth Lloyd Mayer] "As I turned into my driveway [with the harp], I had the thought: This changes everything." . . .* Mayer, *Extraordinary Knowing.* "My single, conscious thought at the time—This changes everything—was inextricably married to a gut feel-

ing, a personal and irrefutable feeling that made me know something had happened. It was the kind of feeling that cognitive neuroscientist Antonio Damasio captured in the title for his hugely influential book on consciousness, *The Feeling of What Happens*. I had to doubt my existing models of reality or I had to doubt myself . . ." Ibid.

WHAT THE DYING SEE

22 *"We all know we are going to die . . . one day," said Teresa Dellar, executive director of the Residence. "This is different." . . .* C. Cornacchia, "What Happens When We Die?" *Montréal Gazette* (February 10, 2007).

23 *David Kessler, former chair of the Hospital Association of Southern California Palliative Care Transition Committee, has observed this phenomenon countless times . . .* See discussion and examples in Kessler, *Visions, Trips and Crowded Rooms*.

23 *[Gail reporting experience to Kessler] "I only know that I've got this trip in front of me, and the time has come" . . .* Ibid.

24 *"I'm going away tonight," the blues singer James Brown told his manager at Christmas 2006 . . .* Jeremy Simmonds, *The Encylopedia of Dead Rock Stars* (Chicago: Chicago Review Press, 2008).

24 *In the most comprehensive, cross-national study of deathbed experiences ever done, the psychologists Karlis Osis and Erlendur Haraldsson confirmed that these intimations of departure even occurred in people who weren't considered by doctors to be terminally unwell . . .* Osis and Haraldsson, *At the Hour of Death*.

24 *[A case reported to Karlis Osis and Erlendur Haraldsson] He said he felt himself for a few seconds to be not in this world but elsewhere . . . 'I am going,' he said, and departed a few minutes later . . .* Ibid., 67.

24 *One paramedic sees this puzzling interior knowledge displayed in his ambulance en route to hospital . . .* http://www.coasttocoastam.com/show/2009/08/28.

25 *"My patient said 'yeah, I'm going to die today,'" a palliative physician recalled in a 2011 report . . .* Shane Sinclair, "Impact of Death and Dying on the Personal Lives and Practices of Palliative and Hospice Care Professionals," *Canadian Medical Association Journal* 183, no.2 (February 8, 2011): 180–87.

25 *"Several medical observers expressed amazement and surprise when confronted with cases in which patients died" . . .* Osis and Haraldsson, *At the Hour of Death*.

26 [*Kathleen Dowling Singh*] *"I do know that she was referring to a process she was aware she was enduring and that that process had, for her, a referential beginning point and end point"* . . . Singh, *The Grace in Dying*. Writes Singh further: "The psychoalchemy of terminal illness seems to allow a natural feeling of safety as death approaches . . . [or] as Sogyal Rinpoche says, [we move toward it] 'as instinctively . . . as a little child running eagerly into its mother's lap, like old friends meeting, or a river flowing into the sea.' It is my observation, by and large, this is so. Jewish wisdom speaks of *devekut*, 'melting into the Divine,'" pp. 222–23.

26 *"Does my wife understand about the passport and ticket?"* . . . Maggie Callanan (Keynote Address, International Association of Near-Death Studies, Raleigh, North Carolina, 2008); see also Callanan and Kelley, *Final Gifts*.

26 *Trained as an ER nurse, she found that in the quieter realm of homes she could observe a distinct pattern of behavior in her dying patients* . . . Ibid.

27 *The famed psychiatrist Elisabeth Kübler-Ross had innovated hospice care during this decade* . . . See, for example, Elisabeth Kübler-Ross, *On Death and Dying* (New York: Macmillan, 1969).

27 *Atlanta psychiatrist Raymond Moody's groundbreaking research on near-death experiences* . . . Moody, *Life After Life*.

28 *A 2009 study of five hospices and nursing homes confirmed that 62 percent of physicians and nurses had encountered what might be called paranormal "death-bed phenomena"* . . . H. Lovelace et al., "Comfort for the Dying: Five Year Retrospective and One Year Prospective Studies of End of Life Experiences," *Archives of Gerontology and Geriatrics* 51, no. 2 (September 2010): 173–79. See also S. Brayne et al., "End of Life Experiences and the Dying Process in a Gloucestershire Nursing Home as Reported by Nurses and Care Assistants," *American Journal of Hospital and Palliative Care* 25, no. 3 (2008): 195–206.

28 *Callanan and Kelley decided to frame what they observed as a distinct state of consciousness, which they dubbed Nearing Death Awareness* . . . Callanan and Kelley, *Final Gifts*.

28 *The book was like a modern* Arte Moriendi . . . "In the Christian Monastic tradition," writes Kathleen Dowling Singh, "the dying enter three phases: separation, liminality, and reincorporation. Liminality was viewed as a period of metamorphosis, the preparatory period for the emergence of the soul on its journey home to Spirit." Singh, *The Grace in Dying*, 216.

28 *I think we have a moral, ethical, and human responsibility to tell our stories, no matter how many times they fall on deaf ears"* . . . Maggie Callanan (in *Vital Signs* 27 no. 3 (2008).

29 *Fifty-four percent of staff in the five-hospice study had patients who experienced a "visit".* . . Lovelace, "Comfort for the Dying."

30 *When I was a student nurse, I remember crossing over with the night shift, and they said very matter-of-factly, 'So-and-so has been chatting with his dead mother for the last five hours, so he'll be off soon'"* . . . Penny Sartori, interview by the author, summer 2012.

30 *[Dianne] Arcangel was taken aback; she had had a near-death experience several years earlier* . . . Arcangel, *Afterlife Encounters*, 119.

30 *Of the ten percent of dying who were conscious in the hour before death, the majority reported seeing such visions* . . . Ibid.

31 *"A cardiac patient, a fifty-six-year-old male whose consciousness was clear, saw the apparition of a woman who had come to take him away"* . . . Ibid.

31 *Searching for other correlations, Osis and Haraldsson determined that less than 10 percent of the patients had high fevers, which can ignite hallucinations* . . . Ibid.

31 *A state called "terminal restlessness"* . . . See, for example, A. Mazzarino-Willett, "Deathbed Phenomena: Its Role in Peaceful Death and Terminal Restlessness," *American Journal of Hospice and Palliative Medicine* 27, no. 2 (2010): 127–33.

32 *As one nurse in a UK hospice study described it, "When they have a high temperature they see things and it's an anxiety-based thing"* . . . S. Brayne, "End of Life Experiences"; see also P. Fenwick et al., "End of Life Experiences and Their Implications for Palliative Care," *International Journal of Environmental Studies* 64, no. 3 (2007): 315–23.

32 *In nearly 80 percent of Osis and Haraldsson's cases, the apparent purpose of the deathbed vision was to accompany or take the patient away* . . . Ibid. Eighty-seven percent of those who saw apparitions with a "take away" purpose died within sixty minutes, as opposed to 46 percent of others, who died after much longer intervals. Also, a majority of patients who saw the "take away" apparitions reacted with moods of elation or serenity, whereas only 7 percent responded to other kinds of hallucinations that way. Ibid., 86.

32 *Pondering other angles, the two psychologists examined what they called "the mirage effect"* . . . Ibid.

---

32 *Patients in distressed or anxious states were less likely to see apparitions than those in calmer moods, their data showed* . . . Ibid. "We clearly found that phenomena . . . are not related to indices of stress, expectations to die or to recover, or the patient's desire to see a person dear to him," p. 190.

33 *The Cane family hastened to reassure Fenwick: "We think that there is probably a medical reason why the dying hallucinate"* . . . Fenwick and Fenwick, *The Art of Dying*.

39 *In the third century BC, the Greek physician Herophilus of Alexandria became the first known man to dissect a human corpse out of pure curiosity* . . . Roach, *Spook*, 68–69.

40 *A radical departure from the classic Egyptian understanding, which held that the soul was in the heart* . . . See, for example, *The Egyptian Book of the Dead: The Book of Going Forth by Day*.

40 *In the first century AD, the Emperor Hadrian asked Rabbi Joshua ben Hananiah to show him the "soul bone" that some Jewish spiritual authorities claimed existed* . . . Roach, *Spook*, 67–68.

40 *Humans drew into themselves with their breath the soul-force of the world, and this "spiritus" flowed through pipes and organs* . . . Ibid., 67.

40 *Descartes believed that spirit flowed to and from the pineal gland through a nervous system that he modeled on a church organ, with microscopic bellows* . . . See, for example, discussion in Zimmer, *Soul Made Flesh*.

41 *The soul, he determined, weighed three-fourths of an ounce* . . . Roach, 110–11.

42 *Briefly, Doris reflected to those in the room that she should, perhaps, stay for the baby's sake. But then she said, 'I can't—I can't stay; if you could see what I do, you would know I can't stay'"* . . . Barrett, *Deathbed Visions*, 13–14.

43 *As a teenager in Lithuania, he had been with an aunt who had deathbed visions* . . . Osis and Haraldsson, *At the Hour of Death*.

43 *A phenomenon that may be related . . . called "terminal lucidity"* . . . M. Nahm and B. Greyson, "Terminal Lucidity in Patients with Chronic Schizophrenia and Dementia: A Survey of the Literature," *Journal of Nervous and Mental Disease* 197, no. 12 (2009): 942–44.

43 *On terminal lucidity* . . . M. Nahm et al., "Terminal Lucidity: A Review and a Case Collection," *Archives of Gerontology and Geriatrics* 55, no. 1 (July 2012); see also Scott Haig, M.D., "The Brain: The Power of Hope," *Time* (January 29, 2007).

43 *Elisabeth Kübler-Ross corresponded with Karlis Osis about their shared obser-*

vations of schizophrenics and stroke patients who suddenly became oriented, direct, and crisp . . . Osis and Haraldsson, *At the Hour of Death.*

44 *Hospice still serves only a portion of the dying population, although it has been steadily increasing* . . . National Hospice and Palliative Care Organization, www.nhpco.org.

44 *Less than 20 percent of British citizens passed away in nonhospital settings in 2008* . . . Fenwick and Fenwick, *The Art of Dying.*

44 *[Of people who die in hospice care] In Canada, it's about 30 percent* . . . "Fact Sheet: Hospice Palliative Care in Canada," Canadian Hospice Palliative Care Association, updated March 2013, http://www.chpca.net/media/7622/fact_sheet_hpc_in_canada_march_2013_final.pdf.

SIGNALS AND WAVES

48 *Research done in Wales, Japan, Australia, and the United States shows that between 40 and 53 percent of the bereaved experience "anomalous cognition" when someone close or connected to them has died* . . . See, for example, percentage of widows and widowers—53 percent: A. M. Greeley, "Hallucinations Among the Widowed," *Sociology of Social Research* 71, no. 4 (1987): 258–65; percentage of general population in U.S.—44 percent: R. A. Kalish and D. K. Reynolds, "Phenomenological Reality and Post-Death Contact," *Journal for the Scientific Study of Religion* (1973): 209–21. For a review of cross-cultural prevalence, see E. Steffen and A. Coyle, "Sense of Presence' Experiences in Bereavement and Their Relationship to Mental Health: A Critical Examination of a Continuing Controversy," in *Mental Health and Anomalous Experience,* ed. C. Murray, *Psychology Research Progress* 3 (Hauppauge, NY: Nova Science Publishers, 2012), 33–56.

48 *Psychiatrists call these experiences "grief hallucinations"* . . . See, for example, Craig Murray and Sheila Payne, *"Grief Hallucinations: A Narrative Review,"* Mental Health, Religion & Culture (2012), ISSN 1469-9737 (submitted).

48 *In 1991, the British neurosurgeon J. M. Small wrote to the medical journal the* Lancet . . . J. M. Small, letter to the editor, "Sixth Sense," *The Lancet* 337 (June 22, 1991): 1550.

49 *In 2012, the psychologist Erlendur Haraldsson reported a comprehensive study he had done on 340 cases of extraordinary encounters around dying and death* . . . Haraldsson, *The Departed Among the Living.*

50 *Cross-cultural surveys show that "about half of all spontaneous [telepathic]*

experiences occur in dreams . . . Research cited in Radin, *Entangled Minds.* "The truth is that to very high levels of confidence we know that information at a distance was successfully perceived in dreams under well-controlled conditions," p. 115.

51 *Odds against chance in a review of spontaneous telepathy studies have been calculated* . . . Ibid., 110. Odds against chance with crisis apparitions were first calculated by Gurney and Myers at Cambridge, who compared the frequency with which people report having hallucinations in a waking, healthy state together with statistics regarding the incidence of death in the United Kingdom, concluding that hallucinations coinciding with death happened too frequently to be attributable to chance. (This research is discussed in Kelly and Kelly, *Irreducible Mind,* 109.)

51 *[Hans Berger] In 1929, Berger unveiled the first technique for "recording the electrical activity of the human brain from the surface of the head"* . . . David Millett, "Hans Berger: From Psychic Energy to the EEG," *Perspectives in Biology and Medicine* 44, no. 4 (Autumn 2001): 522–42.

51 *[Hans Berger] His invention was initially greeted with skepticism* . . . Ibid.

52 *One of the first of the twin studies showed that, when one twin being monitored by EEG was asked to close his or her eyes* . . . *the distant twin's alpha rhythms also increased* . . . Radin, *Entangled Minds,* 18. For more EEG studies, see ibid., pp. 137–40. See also Daryl J. Bem and Charles Honorton, "Does Psi Exist? Replicable Evidence for an Anamolous Process of Information Transfer," *Psychological Bulletin* 115, no. 1 (1994), 4–18; see also, for further discussion of methodology in psi research, http://www.koestler-parapsychology.psy.ed.ac.uk/Psi.html.

52 *In 2013, a study of British twins reported that 60 percent felt they had telepathic exchanges* . . . Göran Brusewitz and Adrian Parker, Department of Psychology, University of Gothenburg, proceedings of the annual meeting of the Parapsychological Association, Viterbo, Italy, August 2013.

52 *The Czech neurophysiologist Jiri Wackermann concluded in 2003 that, "We are facing a phenomenon which is neither easy to dismiss as a methodological failure or a technical artifact, nor understood as to its nature"* . . . Jiri Wackermann, "Dyadic Correlations between Brain Functional States: Present Facts and Future Perspectives," *Mind and Matter* 2, no. 1 (2004): 105–22; see also Jiri Wackermann et al., "Correlations between Electrical Activities of Two Spatially Separated Subjects," *Neuroscience Letters* 336 (2003): 60–64.

52 *In the early 1960s, the University of Virginia psychiatrist Ian Stevenson began to*

investigate what he referred to as "telepathic impressions" . . . Stevenson, *Telepathic Impressions.* For a retrospective analysis of Stevenson's work, see Carlos Alvarado and Nancy Zingrone, "Ian Stevenson and the Modern Study of Spontaneous ESP Experiences," *Journal of Scientific Exploration* 22, no. 1 (2008): 44–53.

53 *Stevenson decided to start by reviewing the first collection of spontaneous telepathy cases known to have been investigated: 165 reports published in the late nineteenth century by the Cambridge scholars Frederic Myers and Edmund Gurney* . . . Gurney, Myers, and Podmore, *Phantasms of the Living.*

53 *Another instance of shared perception* . . . *involved distress—a storm at sea— rather than death* . . . For a full discussion of the Wilmot case, and the way in which skeptical psychologist Susan Blackmore dismissed it by overlooking certain facts, see Kelly and Kelly, *Irreducible Mind.*

54 *Mrs. Wilmot, back in Connecticut, was equally bothered by the impropriety* . . . E. M. Sidgwick, "On the Evidence for Clairvoyance," Proceedings of the Society of Psychical Research 7: 30–99. See discussion in Kelly and Kelly, *Irreducible Mind*, p. 395.

54 *Stevenson found that the cases broke down roughly equally between men and women. Eighty-nine percent occurred when the person was awake, rather than dreaming or dozing* . . . Stevenson, *Telepathic Impressions.*

54 *Oliver Sacks steps lively over this research* . . . *noting simply that "one suspects" the percipients were mostly snoozing* . . . Sacks, *Hallucinations.*

55 *Two-thirds of the gathered cases involved news of an immediate family member. Eighty-two percent pertained to death, or a sudden illness or accident* . . . Stevenson, *Telepathic Impressions.*

55 *"Is it that the communication of joy has no survival value for us, while the communication of distress has?" Stevenson wondered* . . . Ibid.

55 *In assessing thirty-five of his own contemporary cases, Stevenson discovered that a third involved violent death, whereas only 7.7 percent of all deaths in America in that year, 1966, were violent in nature* . . . Ibid.

55 *[Stevenson's] findings were replicated in 2006, when researchers again found a dramatically higher number of abrupt or violent deaths in telepathic impression cases in* . . . Haraldsson, *The Departed Among the Living.*

55 *Stevenson was careful with the cases he, himself, chose to research. He excluded "instances of repeated gloomy forebodings which on one occasion happened to be right"* . . . Stevenson, *Telepathic Impressions.*

55 *In over half the instances, "the percipient's impression drove him or her to take some kind of action apart from merely telling other people about it"* . . . Ibid.

59 *The second factor was possibly, a higher giftedness* . . . Ibid. See also Radin's discussion of traits and factors that seem to correlate with an enhanced capacity to receive telepathic impressions in Radin, *Entangled Minds*.

59 *Janey Acker Hurth, for example, who had sensed her daughter's imminent collision with a car* . . . Stevenson, *Telepathic Impressions*.

59 *Stevenson was struck by how information sometimes gradually came into focus for people. "The percipient's mind," he mused, "may scan the environment for danger to his (or her) loved ones and, when this is detected, 'tune in' and bring more details to the surface of consciousness"* . . . Ibid. This model of what's going on is reminiscent of the U.S.-government-funded Remote Viewing experiments, about which see: Russell Targ, "Why I Am Absolutely Convinced of the Reality of Psychic Abilities, and Why You Should Be, Too" (presentation at the Parapsychological Association Annual Convention, Paris 2010).

60 *Fenwick . . . has amassed more than two thousand accounts of what he calls "deathbed coincidences"* . . . Fenwick and Fenwick, *The Art of Dying. As the psychiatrist Edward Kelly noted, "In addition to the sheer quantity of such cases, the quality of the evidence is such that they cannot be dismissed en masse as unsubstantiated 'anecdotes'"* . . . Kelly and Kelly, *Irreducible Mind*, 407. "Hart et al., for example, analyzed a collection of 165 previously published cases in which one of the criteria for inclusion was a written or oral report made of the experience, or action taken because of the experience, before the corresponding event was learned about normally," ibid. Tyrrell reported that he had found 130 collective cases in the literature," Kelly and Kelly, *Irreducible Mind*, 407, referencing G. N. M Tyrrell, *Apparitions* (London: Duckworth, 1943). "Among cases in which more than one person was present, a third to a half involved collective perception, ibid.

60 *She awoke at 3 a.m. from an intensely vivid dream in which her ex was sitting on the bed* . . . Fenwick and Fenwick, *The Art of Dying*, 63.

60 *Richard Bufton, a college lecturer and commercial diver, was aboard ship when he learned of his grandfather's death via unconventional means* . . . Ibid., 62.

61 *When people hallucinate text, they don't see meaningful messages* . . . Dominic ffytche et al, "Visual Command Hallucinations in a Patient with Pure Alexia," *Journal of Neurology, Neurosurgery and Psychiatry* 75 (2004): 80–86.

---

61 *Some researchers propose that people intuit these death and distress events, garnering the raw information, and then their brains instantly assemble a representation of what they intuited* . . . See, for example, Radin, *Entangled Minds*, and Haraldsson, *The Departed Among the Living*.

62 *A sailor named Raymond Hunter appears to have shared both the pain of his father's illness and the peace that followed* . . . Ibid., 53.

62 *"The syndrome of couvade in which a man imitates the symptoms of his wife's labor pains"* . . . Stevenson, *Telepathic Impressions*, 102. This has subsequently been documented a little more clearly with shared emotional symptoms. See, for example, James F. Paulson et al., "Prenatal and Postpartum Depression in Fathers and Its Association with Maternal Depression: A Meta-Analysis," *Journal of the American Medical Association 303, no. 19* (2010): 1961–69.

62 *A particularly vivid instance comes from an interview conducted by the journalist Paul Hawker in 2010. A woman in her late thirties told him: "I was awoken around 2:00 a.m. by the sound of my heart breaking"* . . . Paul Hawker, *Secret Affairs of the Soul*, 101.

63 *Consider the sailor Raymond Hunter's description: "I remember grabbing my mouth, forcing it open to help me breathe. I was fighting for all I was worth but the pains were now unbearable"* . . . Fenwick and Fenwick, *The Art of Dying*, 54.

63 *The sociologist Glennys Howarth* . . . *researching rare cases of shared illness symptoms across distance* . . . G. Howarth, "Shared Near-Death and Related Illness Experiences: Steps on an Unscheduled Journey," *Journal of Near-Death Studies 20, no. 2* (Winter 2001): 71–85.

64 *"At the age of eighteen," a man named Derek Whitehead told Fenwick he had been working in the Merchant Navy"* . . . Fenwick and Fenwick, *The Art of Dying*, 73.

64 *In the initial phases of critical burn care, the victim must be covered with new skin* . . . Hamilton, *The Scalpel and the Soul*, 71–76.

67 *The Norwegians have a word for this uncanny anticipation of visitors: vardøger* . . . For a fascinating case discussion, see L. David Leiter, "The Vardøgr, Perhaps Another Indicator of the Non-Locality of Consciousness," *Journal of Scientific Exploration 16, no. 4* (2002): 621–34.

68 *We no longer have the old models for understanding them. "Models help us think," writes Berkeley psychiatrist Elizabeth Lloyd Mayer* . . . Mayer, *Extraordinary Knowing*.

---

68 *The French Academy of Sciences in the eighteenth century scoffed at meteorites because: how could rocks fall from the air?* . . . Ibid., 97.

69 *In the late nineteenth century, Hungarian obstetrician Ignaz Semmelweis demonstrated that, if doctors washed their hands before delivering babies, the rates of infection in mothers went down* . . . See, for example, K. Codell Carter and Barbara R. Carter, *Childbed Fever: A Scientific Biography of Ignaz Semmelweis*, 1994.

69 *John Snow was belittled for proposing the existence of germs* . . . See a brief history of his investigation into the causes of cholera here: http://www. ph.ucla.edu/epi/snow/fatherofepidemiology.html.

69 *Hilary Mantel pointed out a few years ago in the* London Review of Books: *"From 1904, the Wright brothers made flights over fields bordered by a main highway and railway line in Ohio* . . . Hilary Mantel, "That Wilting Flower," *London Review of Books* 30, no. 2 (January 24, 2008): 3–6.

69 *A method of measuring telepathy in the lab that came to be known as the "ganzfeld technique"* . . . See discussion in, for example, Daryl J. Bem and Charles Honorton, "Does Psi Exist? Replicable Evidence for an Anamolous Process of Information Transfer," *Psychological Bulletin* 115, no. 1 (1994): 4–18.

69 *"The ordinary waking state is largely driven by sensory awareness, so anything that disrupts that awareness will probably improve psi perception," Dean Radin writes* . . . Radin, *Entangled Minds*.

70 *The receiver, Gail, was told to vocalize her impressions without naming or analyzing them. She was recorded as saying: "Keep feeling like looking up at tall. I'm looking up at tall"* . . . Radin, *Supernormal*.

71 *Between 1974 and 2004, nearly ninety ganzfeld experiments were conducted and published by a number of scientists around the world* . . . Ibid.

72 *Richard Wiseman, a popular skeptic, has said that, "by the standards of any other science"* . . . http://www.dailymail.co.uk/news/article-510762/Could-prove-theory-ALL-psychic.html. See, also, Jime, "Richard Wiseman: The Evidence for ESP meets the Scientific Standards for a Normal Claim," Subversive Thinking, April 5, 2010, http:subversivethinking.blogspot. com/2010/04/richard-wiseman-evidence-for-esp-meets.html.

72 *After sifting through the studies, Rosenthal and a colleague reported that, "the ganzfeld ESP studies regularly meet the basic requirements of sound experimental design"* . . . Bem and Honorton, "Does Psi Exist?"

72 *"It is a scandal," said the Cambridge scholar Henry Sidgwick, "that the dispute*

as to the reality of these phenomena should still be going on" . . . "Presidential Address," *Journal of the Society for Psychical Research* 1: 8.

72 *The Cambridge physicist and Nobel Prize winner Brian D. Josephson told the New York Times in 2003: "There's really strong pressure not to allow these things [psi phenomena] to be talked about in a positive way"* . . . Kenneth Chang, "Do Paranormal Phenomena Exist?" *New York Times*, November 11, 2003.

73 *Harold Puthoff, a physicist at the Stanford Research Institute . . . described this pressure in a series of emails to the psychiatrist Elizabeth Lloyd Mayer* . . . . Mayer, *Extraordinary Knowing*, 114.

73 *In 2011, physicists at Oxford took quantum entanglement to the macro level by briefly entangling two separated diamond crystals that were visible to the eye* . . . J. Matson, "Quantum Entanglement Links Two Diamonds," *Scientific American* (December 1, 2011).

73 *The neuroscientist Michael Persinger, of Laurentian University in Sudbury, Ontario, thinks that he may have demonstrated the entanglement effect between people* . . . Michael Persinger, interview by Alex Tsakiris, Skeptiko.com, Internet radio broadcast, December 16, 2009.

74 *"Quantum theory and a vast body of supporting experiments tell us that some*-thing unaccounted for is connecting otherwise isolated objects" . . . Radin, *Entangled Minds*, 232.

74 *We asked if groups of people brought by circumstances into resonance or coher*-*ence might share a group consciousness that would register in the data from our random devices. The answer was yes"* . . . Roger Nelson and Peter Bancel, "Effects of Mass Consciousness: Changes in Random Data During Global Events," *Explore* 7 (2011): 373–83.

74 *You can see what this looks like, how random data shifts to patterned data on the PEAR Global Consciousness Project website* . . . "The Global Consciousness Project Meaningful Correlations in Random Data," Global Conscious-ness Project, http://noosphere.princeton.edu.

75 *The overall statistics for the project," Nelson wrote, "indicate odds of about 1 in 20 million that the correlation of our data with global events is merely a chance fluctuation"* . . . Roger D. Nelson, "Is the Global Mind Real?" *EdgeScience* 1 (October 2009): 6–8.

75 *The American naturalist William Long published a book called* How Animals Talk, *reporting his observations of wolves in the Nova Scotia wilderness* . . . William Long, *How Animals Talk* (New York: Harper, 1919).

75 *Controlling for scent, the sound of the car on the road, the routine time of day,*
and all familiar sounds, Sheldrake was able to establish that dogs begin to antic-
ipate their owners' arrival regardless of the sensory cues . . . Sheldrake, *Dogs*
That Know When Their Owners Are Coming Home; see also the ensuing
debate between Sheldrake and Richard Wiseman, summarized by Chris
Carter in "Heads I Lose, Tails You Win: How Richard Wiseman Nullifies
Positive Results and What to Do About It," *Journal of the Society for Psychi-*
cal Research 74: 156–67; also, Sheldrake's further discussion with Wiseman
in Sheldrake, *The Science Delusion.*

76 *The South African journalist and author Laurens van der Post described African*
bushmen who knew when a hunting party would return with a kill . . . Laurens
van der Post, *The Lost World of the Kalahari* (New York: Morrow, 1958),
cited in Sheldrake, *The Science Delusion.*

76 *The Iroquois of North America refer to such communication as using "the long*
body" . . . William Roll, "Psi and the Long Body," *Australian Journal of*
Parapsychology 8, no. 1 (2008).

77 *"It is altogether probable that important unrecognized exchanges of feelings*
through extrasensory processes are occurring all the time" . . . Stevenson, *Tele-*
pathic Impressions.

77 *Dean Radin offers one theory about what might be going on to facilitate our*
gleanings of danger . . . Radin, *Entangled Minds*, 265.

78 *In 2010, Cornell Psychologist Daryl Bem published research* . . . "Feeling
the Future: Experimental Evidence for Anomalous Retroactive Influences on Cognition and Affect," *Journal of Personality and Social Psychology*
100 (2011): 407–25.

79 *As one of his peer reviewers, Joachim Krueger of Brown University, put it: "My*
personal view is that this is ridiculous and can't be true . . . Peter Aldhous,
"Is This Evidence We Can See the Future?" *New Scientist* (November 11,
2010).

80 *"The ability to anticipate and thereby to avoid danger confers an obvious evolu-*
tionary advantage" . . . Ibid., 51. This idea is also discussed in G. Hitchman
et al., "A Re-Examination of Non-Intentional Precognition with Open-
ness to Experience, Creativity, Psi Beliefs and Luck Aeliefs As Predictors
of Success," (presentation at the Parapsychological Association Annual
Convention, Curitiba, Brazil 2011): "Stanford's Psi-Mediated Instrumental
Response model predicts that psi can operate in the absence of conscious
awareness, facilitating advantageous outcomes for the organism by trig-

gering preexisting behaviors in response to opportunities or threats in the environment."

80 *Bem was building on the insights of the psychologist Hans Eysenck, who argued that "psi might be a primitive form of perception antedating cortical developments in the course of evolution"* . . . H. J. Eysenck, "Personality and Extra-Sensory Perception," *Journal of the Society for Psychical Research* 44 (1966): 55–71.

80 *The face might trigger a physiological reaction, such as changes in skin temperature (blushing or blanching) or accelerated heart rate* . . . Radin, *Supernormal*, 165.

81 *Bengalese finches show alarm up to nine seconds before the video monitors next to their cages actually display a horseshoe whip snake seeming to approach them* . . . Ibid., 176.

81 *Other experiments in presentiment continue apace all over the world, studying (in no particular order) college students, earthworms, zebra finches, and Zen meditators* . . . See: F. Alvarez, "Anticipatory Alarm Behaviour in Bengalese Finches," *Journal of Scientific Exploration* 24, no. 4 (2010); D. Radin et al., "Electrocortical Activity Prior to Unpredictable Stimuli in Meditators and Non-Meditators," *Explore* 7 (2011): 286–99; D. Bem, "Feeling the Future"; C. Wildey, "Impulse Response of Biological Systems," Department of Electrical Engineering, University of Texas, 2001.

81 *Studies dating back to 1978 established a small but statistically significant incidence of precognition in experiments without methodological flaw* . . . Julia Mossbridge, Patrizio Tressoldi, and Jessica Utts, "Predictive Physiological Anticipation Preceding Seemingly Unpredictable Atimuli: A Meta-Analysis," *Frontiers in Psychology* (October 2012).

81 *Before agreeing to publish Mossbridge's article . . . one of the peer reviewers for the journal requested that a line be inserted to say that this phenomenon was due to natural physical processes, if not yet determined* . . . Email conversation between Mossbridge and author. October 20, 2013.

81 *Lev and Sveta were walking to the lake across a field that skirted the forest* . . . Figes, *Just Send Me Word*, 282.

ASTRAL FATHER

84 *"Steve grinned as he released my hands, then put his warm hand over my mouth as I started to shout my happiness"* . . . Bird's memoir, *Warm Hands*, quoted in Tim Cook, "Grave Beliefs: Stories of the Supernatural and the

Uncanny among Canada's Great War Trench Soldiers," *Journal of Military History* 77 (April 2013): 523–56; see also: Roper, *The Secret Battle.*

84 *George Maxwell would write, for example, of being separated from his platoon in no-man's-land on a lightless evening* . . . Cook, "Grave Beliefs."

85 *"Who is the third who walks always beside you?"* . . . Eliot, "What the Thunder Said," section V in *The Waste Land.*

85 *Later . . . he refused to surrender such a sublime experience to ridicule: "None of us care to speak about that"* . . . Ernest Shackleton, interview with the *Daily Telegraph* (London), February 1, 1922, cited by John Geiger in *The Third Man Factor.*

86 *Polar explorers had been tricked into the illusion by monotony and sensory deprivation—a world of white* . . . D. Chan et al., "But Who Is That on the Other Side of You? Extracampine Hallucinations Revisited," *Lancet*, December 21, 2002.

86 *"We had all three been sensible of his presence throughout the most trying part of the night"* . . . Geiger, *Third Man Factor*, 61. Geiger's book is a must-read for anyone who wants to read about this subject in greater depth.

86 *Three of the World Trade Center survivors on 9/11 later claimed that they were guided to safety by sensed presences* . . . Ibid., 1.

87 *"She was wearing a headscarf and a long dress. She was shadowy and two-dimensional, like a silhouette"* . . . Coffey, *Explorers of the Infinite.*

87 *Lead researcher Olaf Blanke . . . implicated this area of the brain because it "integrates sensory input into a cohesive picture"* . . . O. Blanke, A. Shahar, et al., "Induction of an Illusory Shadow Person," *Nature* 443 (September 21, 2006); see also O. Blanke and T. Metzinger, "Full-Body Illusions and Minimal Phenomenal Selfhood," *Trends in Cognitive Science* 13 (2009): 7–13; see also M. Persinger, "Experimental Facilitation of the Sensed Presence: Possible Intercalation between the Hemispheres Induced by Complex Magnetic Fields," *Journal of Nervous and Mental Disease* 190, no. 8 (2002): 533–41.

88 *"Explanations of the sensed presence phenomenon abound, which is paradoxical, given the paucity of systematic research on the subject"* . . . Peter Suedfeld and John Geiger., "The Sensed Presence as a Coping Resource in Extreme Environments," in *Miracles: God, Science, and Psychology in the Paranormal; Parapsychological Perspectives*, ed. J. Harold Ellens, vol. 3 (Westport, CT: Praeger Publishers, 2008), 1–15.

89 *The American climber James Sevigny was so severely injured by an avalanche in*

Banff National Park in 1983 that he could barely move . . . Geiger, *The Third Man Factor*, 7–10.

90 *Joshua Slocum, the first man to sail solo around the world, reported that, in the midst of a battering storm,* . . . Ibid., 49–52.

90 *Edith Stearns, a contemporary of the more famous female flier Amelia Earhart, came to expect a sensed presence on her juddering flights in the 1930s and '40s* . . . Ibid., 92.

91 *The pilot Brian Shoemaker, disoriented in his H-34 helicopter during an Antarctic storm* . . . Ibid., 94–96.

91 *Wrote the shipwreck survivor Ensio Tiira, who had been adrift on a raft for thirty days, "I'd lost all sense of a second person being in the raft"* . . . Ibid., 199–202.

91 *Hour upon hour this companion watcher, as I call him, peers out at me through the curtain of snow," he later wrote* . . . Taylor, *The Breach*.

92 *It puts me in mind of a poem by the American poet Mary Oliver about the dying visions of William Blake* . . . Oliver, "Blake Dying," in *A Thousand Mornings*, 29.

92 *"Something beyond our understanding occurs in the genesis of qualia," the neurologist Oliver Sacks has written* . . . Oliver Sacks, "In the River of Consciousness," *New York Review of Books*, January 15, 2004.

93 *"There is a fundamental explanatory gap between brain activation and conscious experience,"* . . . *Alexander Morelos said* . . . Alexander Morelos (presentation at the 2012 conference on the "Science of Consciousness," at the University of Arizona, Tucson, Arizona).

94 *Wrote Sacks: "Philosophers argue endlessly over how these transformations occur"* . . . Sacks, "In the River of Consciousness."

94 *The neurophysiologist Richard J. Davidson at the University of Wisconsin-Madison* . . . *has been studying the brain waves of meditating monks* . . . See, for example, the lab website: http://psyphz.psych.wisc.edu.

94 *"It is unknown at present whether hallucinations are generated by similar mechanisms in patients and in healthy people"* . . . Patricia Boksa, "On the Neurobiology of Hallucinations," *Journal of Psychiatry and Neuroscience* 34, no. 4 (July 2009): 260–2.

95 *Britain's National Health Service gave King's College, London, research funds in 2012 to orchestrate collaborative research* . . . King's College, London, news release, posted to their website, June 29, 2012.

95 *As the physicist Harold Puthoff said, it isn't what we know that gets in our way, it's what we believe* . . . Mayer, *Extraordinary Knowing*.

95 *Martha Farah, director of the Center for Neuroscience and Society at the University of Pennsylvania, has said it best: "We should cultivate a certain epistemological modesty," she told colleagues* . . . Martha Farah, "The More We Understand, the Less We Know" (presentation at the Nour Foundation, 2009 Symposia Series).

96 *"Infused with a sense of comfort, warmth, and strength, " she told the journalist John Geiger* . . . Geiger, *The Third Man Factor.* See Bancroft's full account in Liv Arnesen and Ann Bancroft, *No Horizon Is So Far: Two Women and Their Historic Journey Across Antarctica* (Cambridge, MA: Da Capo Press, 2003).

96 *The astronaut Jerry Linenger was working on the space station Mir in 1997 when he became aware of the presence of his father, who had died in 1990* . . . Geiger, *The Third Man Factor*, 102.

96 *In the grief literature, sensing the presence of someone deceased has been defined as "clearly seeing a figure of a human form, someone who was not physically present at that moment"* . . . Steffen, "'Sense of Presence' Experiences in Bereavement and Their Relationship to Mental Health."

96 *Visual perception of the presence seems to be the rarest. Only about 5 percent in one study actually saw the deceased* . . . Barbato, M. "Parapsychological Phenomena." Also the statistical breakdown in Haraldsson, *The Departed Among the Living.*

97 *In a 2006 study, it was found that* . . . Haraldsson, *The Departed Among the Living.*

97 *Sigmund Freud's . . . described healthy recovery from loss as the successful severing of ties* . . . Freud's essay *Mourning and Melancholia,* as discussed in Steffen, "'Sense of Presence' Experiences in Bereavement and Their Relationship to Mental Health."

97 *Those who sensed presences were, in Freud's view, "clinging to the object through the medium of a hallucinatory wishful psychosis"* . . . Ibid.

98 *"The hearing or seeing of a close, recently deceased friend or relative is not a mental disorder," explains one Intro to Psychiatry book* . . . Pridmore, Saxby, *Download of Psychiatry,* ebook prepared for University of Tasmania School of Medicine, http://eprints.utas.edu.au/287/.

98 *In a study of surviving AIDS partners in San Francisco, 22 percent were still sensing or seeing their beloved three and four years later* . . . T. A. Richards, "Spiritual Resources Following a Partner's Death from AIDS," in *Meaning*

Reconstruction and the Experience of Loss, ed. Richard A. Neimeyer (Washington, DC: American Psychological Association), 173–90.

98 *Continues the textbook: "They are comforting and benign. Perhaps they have a role in helping the individual adjust to the loss"* . . . Pridmore, *Download of Psychiatry.*

99 *In 2008, psychiatrist Vaughan Bell wrote* . . . Vaughan Bell, "Ghost Stories: Visits from the Deceased," *Scientific American* (December 2, 2008).

99 *Here is an account from a lawyer, interviewed by the psychologist Erlendur Haraldsson* . . . Haraldsson, *The Departed Among the Living.*

102 *Surveys of people's reactions to grief counseling show, unsurprisingly, that they often feel "unaccepted, abnormal, not understood"* . . . E. Steffen, "Sense of Presence Experiences and Meaning-Making in Bereavement: A Qualitative Analysis," *Death Studies* 35 (2011): 579–609.

102 *A study of sensed presence experiences in Norwegian widows* . . . T.C. Lindstrom. "Experiencing the Presence of the Dead: Discrepancies in the 'Sensing Experience' and Their Psychological Concomitants," *Omega: Journal of Death & Dying 31 (1995)* 11–21.

102 *Moved and astonished by the uncanny experiences he repeatedly encountered in his patients* . . . Barbato, "Parapsychological Phenomena at time of death." *Journal of Palliative Care* (Summer 1999): 15, 2.

102 *Here is an account from the Scandinavian writer Johan Kuld, about an experience he had shortly after the death of his wife* . . . Haraldsson, *The Departed Among the Living,* 113.

103 *From Rome through to nineteenth-century Iceland, the living shared their pastures and roads with those who'd predeceased them* . . . Lecouteux, *The Return of the Dead.*

103 *To keep them at bay, in some societies bodies were decapitated before burial or tied firmly to the ground* . . . See discussion in Taylor, *The Buried Soul.*

103 *"The world was haunted, by the dead transformed into spirits passed on to another state, living another life in permanent conjunction with 'contemporary' humans"* . . . Lecouteux, *The Return of the Dead.*

104 *The English historian William of Newburgh wrote in the 1190s* . . . William of Newburgh, *History of English Affairs,* book 5, ch. 24 (New York: Fordham University Press, 2000), http://www.fordham.edu/halsall/basis/williamof newburgh-intro.asp. Lecouteux agrees with Newburgh: "Revenants were no cause for surprise to the Germanic peoples; they fit perfectly within their

mind-sets, their place has not been usurped, and we cannot dismiss these stories as 'old wives tales.' The roots of the belief are too deep," Ibid, 155.

104 *"Augustine posed the problem of perception"* . . . Ibid.

104 *Chaz Ebert, the Chicago lawyer and widow of the film critic Roger Ebert, awoke the night that her first husband's father died* . . . Chaz Ebert, interview with Studs Turkel, in his oral history, *Will the Circle Be Unbroken? Reflections on Death, Rebirth and a Hunger for Faith.*

105 *Dean Radin writes: "That vision is a construction from your memory and imagination, similar to a waking dream, except that the stimulus for the image is occurring somewhere, or some-when, else"* . . . Radin, *Entangled Minds.* Leaving aside for a moment the origin of the signal, the idea that we can construct visual meaning is now commonly understood within the field of perceptual studies. The most elementary example is this: "Yuo mgiht aslo be srupsired to fnid taht yuo can raed tihs ntoe wtihuot mcuh trouble." (See further discussion in Kelly and Kelly, *Irreducible Mind.*)

105 *Perhaps, speculates the psychologist Erlendor Haraldsson, "The deceased moulds the perception in the mind of the living person"* . . . Haraldsson, *The Departed Among the Living.*

105 *Pope Gregory the Great (540–604) pursued the quandary of apparitions down a path that would ultimately contribute a little to the establishment of Purgatory* . . . Lecouteux, *The Return of the Dead.*

107 *Approximately 30 percent of the population has had a simple episode of waking up feeling briefly paralyzed, but encounters with the malignant sensed presence happen to 3 to 6 percent* . . . A. Cheyne, "Relations among Hypnagogic and Hypnopompic Experiences Associated with Sleep Paralysis," *Journal of Sleep Research* 8 (1999): 313–17.

110 *"We have been especially struck with the frequency that the specific term 'evil' is applied to this presence"* . . . A. Cheyne, http://watarts.uwaterloo. ca/~acheyne.

110 *I literally fear for my soul* . . . Ibid.

110 *Research at Fukushima University in Japan has established that people experiencing sleep paralysis show an EEG signature that mixes wakefulness with REM sleep* . . . . Warren, *The Head Trip.*

110 *"Hypnogogic experiences are a bit of a mystery"* . . . Ibid.

111 *There is a medieval account from Sweden where the presence is said to sound like a cloth sack dragging across the floor* . . . David Hufford, interviewed for the

2008 television documentary *Your Worst Nightmare: Supernatural Assault.*
Observations confirmed in email correspondence with the author in 2013.
See also D. Hufford, "Visionary Experiences in an Enchanted World,"
Anthropology and Humanism 35, no. 2 (2010): 142–58.

111   *The Navaho have heard moccasins shuffling; patients have heard stockinged feet*
on hospital carpets . . . Ibid.

111   *Then there is the dread beyond words: "The greatest primal terrors that I have*
ever witnessed: character-forming stuff" . . . Cheyne et al.

111   *"These attacks leave me shuddering and crying," says another. "Sometimes I'm*
so scared I get sick to my stomach" . . . Ibid.

111   *"The complex pattern of sleep paralysis, including the evil presence, the shuffling*
footsteps and a host of other details, does not arise from culture" . . . Hufford,
Your Worst Nightmare.

111   *Thus we have the succubus and incubus of medieval Europe, or the "sitting*
ghost" of China . . . Hufford, *The Terror That Comes in the Night.*

111   *[It] is the origin of the word "haggard" which began as "hag-rid," or being rid-*
den by the Hag . . . Ibid.

112   *Similarly, "nightmare" comes from the Anglo-Saxon word "mare," or "crusher"*
. . . Ibid.

112   *"The neurophysiology of the paralysis is very well understood, we know the bio-*
chemistry and we know the pathways" . . . Hufford, *Your Worst Nightmare.*

113   *Perceiving a spiritual being, whether loving or cruel, has become, Hufford pro-*
poses, "an illegal experience" . . . Ibid.

113   *One influential study of Japanese widows* . . . J. Yamamoto et al., "Mourn-
ing in Japan," *American Journal of Psychiatry* 125 (1969): 1660–65. See also
A. Grimbly, "Bereavement among Elderly People: Grief Reactions,
Post-Bereavement Hallucinations and Quality of Life," *Acta Psychiatry*
Scandinavia 87 (1983): 72–80: See also W. Rees, "The Hallucinations of
Widowhood," *British Medical Journal* 4 (1971): 37–41; Steffen, "Sense of Pres-
ence Experiences and Meaning-Making in Bereavement"; L. E. LeGrand,
"The Nature and Therapeutic Implications of the Extraordinary Experi-
ences of the Bereaved," *Journal of Near-Death Studies* 24 (2005): 3–20.

114   *"There is a danger that in objectifying or analyzing an experience we may lose*
sight of its significance for the bereaved or the dying" . . . Barbato, "Parasycho-
logical Phenomena," 30–37.

114   *"Where is the wisdom we have lost in knowledge?" T. S. Eliot asked in his poetry*
almost a hundred years ago now . . . Eliot, "Choruses from 'the Rock.'"

Notes

BE STILL

116 *The dying experienced an "elevation in mood" right before they passed away* . . . Osis and Haraldsson, *At the Hour of Death.*

119 *Research shows that when people are confronted by their own demise* . . . R. Noyes and R. Kletti, "Depersonalization in response to Life-threatening Danger." *Comprehensive Psychiatry* (July-August 1977): 375–384

120 *In the Atlanta-based cardiologist Michael Sabom's 2004 study of accident survivors* . . . M. Sabom, "The Acute Dying Experience," *Journal of Near-Death Studies* 26, no. 3 (2008).

121 *In 1893, the Swiss geologist Albert Heim published his exploration of the emotional state of mountain climbers who had fallen in the Alps* . . . Heim, "Remarks on Fatal Falls."

121 *A U.S. marine in a 15,000-foot free fall into the Pacific Ocean due to a failed parachute recalled this startling shift in his perspective* . . . Sabom, "The Acute Dying Experience."

122 *When he "returned" to his body after being revived by the lunch-thieving guide, who had now found him, "my grief was measureless"* . . . Cited in Tompkins, *The Modern Book of the Dead.*

122 *The Psychoanalyst Oskar Pfister, a contemporary of Freud's, read some of these mountaineering accounts and called the experience of Heim's climbers "shock thoughts"* . . . O. Pfsiter, "Shocken und Shockphantasien be hochster Todesgefahr," *Zeitschrift fur Psychoanalyse* 16 (1930): 430–55.

122 *The psychiatrist Russell Noyes of the University of Iowa researched such experiences in the 1970s and concluded that "depersonalization as a defense against the threat of extreme danger or its associated anxiety" was the best available explanation* . . . R. Kletti and R. Noyes, "Mental States in Mortal Danger," *Essence* 5, no. 1 (1981): 5–20.

123 *"Considered together," Sabom writes, "these findings present an interesting paradox: the more terrifying and traumatic an accident may appear, the more peaceful and painless the accident may be experienced"* . . . Sabom, "The Acute Dying Experience."

125 *"Contemporary testimony bears a striking resemblance to medieval narratives"* . . . Zaleski, *Other World Journeys.*

125 *Spoke Zoraster: "Righteous souls will enter heaven, and there they will have a vision of God, who is depicted as pure light"* . . . F. Masumian, "World Religions and Near-Death Experiences," in Holden, *The Handbook of Near-Death Experiences.*

125 *"The Lord," says Isaiah in the Torah, "will be your everlasting light"* . . . Isaiah 60:19.

125 *In Hinduism: "Place me in that deathless, undecaying world / Wherein the light of heaven is set"* . . . Ibid. Referencing R. Griffith, *Hymns of the Rigveda* (Benares, India: Medical Hall Press, 1926).

125 *Jesus: "I am the light of the world"* . . . John 8:12.

125 In Buddhism: Clear Light, and Infinite Light . . . Masumian, "World Religions."

125 *Said the Bahai prophet Abdu'l-Baha . . . "That divine world is manifestly a world of lights"* . . . Adbu'l-Baha (lecture, 309 West Seventy-Eighth Street, New York, NY, July 6, 1912), US Bahá'í Publishing Trust, 2nd ed. (1982), 470.

125 *The sixth-century pope Gregory was obsessed with collecting firsthand accounts of spiritual encounters and reported what he'd heard from witnesses of the light* . . . Zaleski, *Other World Journeys.*

126 *About a thousand years later, the accomplished and respected sixteenth-century Spanish abbess and mystic Teresa of Ávila* . . . From *The Book of Her Life,* excerpted in Mandelker, *Pilgrim Souls,* 300.

126 *Six hundred years after that, hospital patient Monique Hennequin described it to her cardiologist* . . . Lommel, *Consciousness Beyond Life,* 203–21.

126 *Another patient of van Lommel's exclaimed, "Too much! It's simply too much for human words"* . . . Ibid.

127 *When Beverly Brodsky was twenty, she lost half the skin on her face in a traumatizing motorcycle accident* . . . Cited in K. Ring and E. Valarino, *Lessons from the Light: What We Can Learn from the Near-Death Experience* (Needham, MA: Moment Point Press, 2006), 236.

128 *In the 1920s, the German theologian Rudolf Otto tried to grapple with the impossibility of understanding how light could, at the same time be love* . . . Otto, *The Idea of the Holy.*

129 *The numinous, he wrote, "is beyond our apprehension because, in it, we come upon something 'wholly other'"* . . . Ibid.

130 *"Oh, that I could tell you what the heart feels, how it burns and is consumed inwardly!" said the sixteenth-century Italian mystic Catherine of Genoa* . . . Ibid.

130 *In his recent book on hallucinations* . . . Sacks, *Hallucinations.*

130 *One woman described the day after her mother died* . . . Haraldsson, *The Departed Among the Living.*

130 *"Everything turns upon the character of this overpowering might," Otto wrote* . . . Otto, *The Idea of the Holy.*

131 *"Patients reacted to otherworldly visions with otherworldly feelings"* . . . Osis and Haraldsson, *The Hour of Our Death.*

131 *Peak spiritual experiences in Buddhist meditators and Franciscan nuns* . . . Newberg and d'Aquili, *Why God Won't Go Away.*

132 *The scans revealed a "sharp reduction" in blood flow to the posterior superior parietal lobe, which is the part of the brain that orients us in space* . . . Ibid., 119.

132 *When it is damaged by a stroke, for example, it becomes impossible for a person to tell where he ends and the room around him begins* . . . Bolte-Taylor, *My Stroke of Insight.*

132 *Experienced meditators are able to quiet or disable this brain region temporarily, preventing sensory data from reaching it and enabling them to experience what has been called the unio mystica.* . . . Newberg and d'Aquili, *Why God Won't Go Away*, 126.

132 *The fourteenth-century German mystic Johannes Tauler describes the feeling as being as if the soul. . . . becomes "sunk and lost in the abyss of the Deity"* . . . Underhill, *Mysticism.*

133 *The German mystic Meister Eckhart in the fourteenth century* . . . Ibid.

133 *At this point in our research," Newberg wrote, "science had brought us as far as it could"* . . . Newberg and d'Aquili, *Why God Won't Go Way*, 127.

133 *Subsequent research by the Montréal neuroscientist Mario Beauregard, imaging the brains of Carmelite nuns* . . . Beauregard, *The Spiritual Brain.*

134 *Nurse Maggie Callanan commented on this recently in a radio interview: "They, on some level, understand that it's odd, that they have one foot in two worlds* . . . Maggie Callanan, interview by Joan Herrmann, *Change Your Attitude . . . Change Your Life*, 2011.

DEEPER: WHAT THE NDEs TELL US

140 *"In the light without a body I could handle that level of love because I had left the physical side of emotion behind* . . . Bennett, *Voyage of Purpose.*

140 *Journalist Ptolomy Tompkins described this from the Buddhist perspective* . . . Tompkins, *The Modern Book of the Dead.*

141 *As Sogyal Rinpoche writes* . . . *"The memory of your inner nature, with all its splendor and confidence, begins to return to you"* . . . Rinpoche, *The Tibetan Book of Living and Dying*, 124.

141 *Near-death experiences have been reported by 12 to 18 percent of the American population* . . . See discussions about prevalence and incidence in Lommel, *Consciousness Beyond Life,* 9, and Parnia, *Erasing Death,* 173.

141 *Over 90 percent of cardiac arrest survivors have no memory of what happened before and after their resuscitation due to brain trauma* . . . Parnia, *Erasing Death.*

142 *The psychologist Carlos Alvarado, who has written extensively on the subject of NDEs, fixes the known incidence rate at 17 percent* . . . See discussion in N. Zingrone and C. Alvarado, "Pleasurable Western Adult Near-Death Experiences: Features, Circumstances and Incidence," in Holden, *The Handbook of Near-Death Experiences,* 34.

142 *The Near-Death Experience Scale* . . . B. Greyson, "The Near-Death Experience Scale: Construction, Reliability and Validity," *Journal of Nervous Mental Disorders* 171 (1983): 369–75.

143 *One man, a Californian arms dealer who had sold weapons in Latin America, had a near-death experience when he was struck by lightning in 1984* . . . Story of Dannion Brinkley, *Act of God,* directed by Jennifer Baichwal (Foundry Films 2009).

143 *Small kindnesses also appear in the life review experience, driving home the point that love and compassion are "the absolute coin of the realm"* . . . Eben Alexander (public talk, Syracuse, New York, May 15, 2013).

145 *"In the chambers of the sea, by sea-girls wreathed in seaweed, red and brown"* . . . Eliot, "The Love Song of J. Alfred Prufrock," in *Prufrock and Other Observations.*

145 *The* Guardian *newspaper arrives in my mail featuring the headline: "Near-Death Experiences: The Brain's Last Hurrah"* . . . Anil Seth, *Guardian,* August, 28, 2013, p. 20.

146 *Citing a new study* . . . *where induced heart attacks in rats that demonstrates continuing electrical activity in the rodent brain for thirty seconds postmortem* . . . "Electrical Signatures of Consciousness in the Dying Brain," University of Michigan Heath System, August 12, 2013, http://www.uofmhealth.org/news/archive/201308/electrical-signatures-consciousness-dying-brain.

146 *"The most important objection to the adequacy of all [these] theories," writes the psychiatrist Bruce Greyson* . . . Bruce Greyson, "Cosmological Implications of Near-Death Experiences," *Journal of Cosmology* 14 (2011): 4684–96. It has been argued, by, for instance, John Horgan, author of *Rational Mysticism,*

that brain imaging can identify the wiring but doesn't explain the source current for these perceptions. We know that people enter into a different realm of consciousness, but not why.

147 *"I remember joy pouring out of my being"* . . . Jayne Smith (panel presentation, annual meeting of the International Association of Near-Death Studies, Raleigh, North Carolina, 2008).

148 *Unconsciousness during anesthesia, as measured by EEG, is associated with an immense quieting of brain activity* . . . Greyson, "Cosmological Implications." Further, "anecdotal reports that adequately anesthetized patients retain a significant capacity to be aware of or respond to their environment in more than rudimentary ways—let alone hear and understand—have not been substantiated by controlled studies," M. Ghoneim and R. Block, "Learning and Consciousness During General Anesthesia," *Anesthesiology* 76 (1992): 279–305.

148 *The closer that NDEers are to actual death—such as being clinically brain-dead or in cardiac arrest—the more likely they are to encounter deceased people* . . . E. W. Kelly, "Near-Death Experiences with Reports of Meeting Deceased People," *Death Studies* 25 (2001): 229–49.

149 *A nine-year-old boy with meningitis awakened from a coma and told his parents that he had seen his teenaged sister on the other side—she had in fact just died in a car accident, unbeknownst to the boy and his parents* . . . Ibid. See also B. Greyson, "Seeing Deceased Persons Not Known to Have Died: 'Peak in Darien' Experiences,'" *Anthropology and Humanism* 35 (2010): 159–71.

149 *Children, in particular, will sometimes spontaneously recognize photographs of dead relatives in the family album after their NDE, as detailed in several studies* . . . See, for example, B. Greyson, "Cosmological Implications," 4684–96; B. Greyson, "Seeing Deceased Persons not Known to Have Died: 'Peak in Darien' experiences"; C. Sutherland, "Trailing Clouds of Glory: The Near-Death Experiences of Children and Teens," in Holden, *The Handbook of Near-Death Experiences*, 87–108.

149 *"Dutch cardiologist Pim van Lommel reports that people who had deep NDEs* . . . Lommel, *Consciousness Beyond Life*, 145.

149 *About 56 percent of Westerners who have NDEs experience themselves going out of body* . . . Carter, *Science and the Near-Death Experience*.

149 *A child who sustained a major head injury in a car accident spent months in a coma before returning to waking consciousness* . . . Sutherland, "Trailing Clouds of Glory."

150 *Here is a Maori account: "I became so ill that my spirit actually passed out of my body"* . . . A. Kellehear, "Census of Non-Western Near-Death Experiences to 2005: Observations and Critical Reflections," in Holden, *Handbook of Near-Death Experiences,* 135–58.

150 *There are accounts in Hawaii of walking toward Pele's Pit, a volcano into which dead souls hurl themselves* . . . Ibid.

151 *A 2013 study of severely brain-injured patients suffering posttraumatic amnesia in Guangzhou, China, uncovered three NDEs out of eighty-six patients* . . . Yongmei Hou et al., "Infrequent Near Death Experiences in Severe Brain Injury Survivors: A Quantitative and Qualitative Study," *Annals of Indian Academy of Neurology* 16, no. 1 (2013): 75–81. (Commenting on what is known to date about NDEs, the authors state, "Although no theory to date has been able to even come near to providing a complete explanation of NDE, a wide range of physiological processes have been targeted for this purpose . . . At best, these physiological findings can be stated as correlates of NDE, rather than its causative biological underpinning.")

151 *In India, people often encounter the spiritual equivalent of irritated bureaucrats, telling them they're the wrong one* . . . Osis and Haraldsson, *At the Hour of Death.*

151 *"Clearly, NDErs meet an assortment of social beings, and their previous experiences shape their interpretation"* . . . Kellehear, *Census of Non-Western Near Death Experiences.*

151 *In Carl Sagan's science-fiction novel* Contact, *Dr. Ellie Arroway travels to meet some unknown alien intelligence* . . . Sagan, *Contact.*

152 *NDEs "seem to be unique, unrivalled memories"* . . . Marie Thonnard et al., "Characteristics of Near-Death Experiences Memories as Compared to Real and Imagined Events Memories," *PLoS ONE* 8, no. 3 (March 2013).

152 *Neurologists call them flashbulb memories, in that a "highly emotional, personally important and surprising event can benefit from a preferential encoding that makes them more detailed and longer-lasting"* . . . Ibid.

152 *There is no evidence that NDEs become confabulated over time* . . . B. Greyson, "Consistency of Near-Death Experience Accounts over Two Decades: Are Reports Embellished over Time?" *Resuscitation* 73 (2007): 407–11.

152 *"Interestingly," the Belgian researchers noted, "NDE memories in this study [also] contained more characteristics than coma memories, suggesting that what makes the NDEs 'unique' is not being 'near-death' but rather the perception of the experience itself"* . . . Ibid.

153 *Hallucinations related to morphine and other palliative drugs "tended to be random and non-specific"* . . . Penny Sartori, telephone interview with author. See also P. Sartori, "A Prospective Study of NDEs in an Intensive Therapy Unit," *Christian Parapsychologist* 16, no. 2 (2004): 34–40.

153 *"When hallucinating individuals return to normal consciousness, they immediately recognize the fragmented and dreamlike nature of their hallucinatory interlude, and understand it was all a mistake of the mind"* . . . Newburgh and D'Aquili, *Why God Won't Go Away*, 112.

153 *Wrote Teresa of Ávila in sixteenth-century Spain: "God visits the soul in a way that prevents it doubting when it comes to itself that it has been in God and God in it"* . . . Underhill, *Mysticism*.

153 *Carl Jung wrote, after his near-death experience during a heart attack in Switzerland in 1944* . . . Jung, *Memories, Dreams, Reflections*.

154 *In the aftermath of their NDEs, people describe the world around them as flat, dull, two-dimensional* . . . *Back to Life: Six Challenges Faced by Near-Death Experiencers* (presented by psychologist Yolaine Stout at IANDS conference, 2008).

154 *I could see the life energy in my surroundings. There was an aura of light around all the plants and rocks in the planting beds"* . . . Bennett, *Voyage of Purpose*.

155 *The New York journalist Maureen Seaberg* . . . *calls it "a God hangover"* . . . Seaberg, *Tasting the Universe*.

155 *When Moses was on Mt. Sinai* . . . ibid.

156 *Listening to the tapes, the Poles couldn't understand Carsolio and he couldn't understand them* . . . Coffey, *Explorers of the Infinite*.

156 *The thirteenth-century Jewish mystic Abraham Abulafia engaged in prayers that caused the letters of the alphabet to come across to him as musical notes* . . . Seaberg, *Tasting the Universe*.

156 *Research at the University of California, Irvine, by the psychiatrist Roger Walsh determined that 86 percent of advanced meditation practitioners experienced synesthesia* . . . R. Walsh, "Can Synaesthesia Be Cultivated? Indications from Surveys of Meditators," *Journal of Consciousness Studies* 12 (2005).

156 *Said David Bennett, of his time beneath the sea: "My perceptions were unbounded"* . . . David Bennett, interview by author.

156 *A study of people who had near-death experiences . . . found that 83 percent had these kinds of synesthesic aftereffects* . . . See discussion in Perera, Jagadheesan, and Peake, *Making Sense of Near-Death Experiences*, 72.

156 *Meanwhile, Brown University's Willoughby Britton discovered that people*

showed differences in their left temporal lobe activity after an NDE . . . W. B. Britton and R. R. Bootzin, "Near-Death Experiences and the Temporal Lobe," *American Psychological Society* 15, no. 4 (2004): 254–58.

157 *Synesthesia has been associated with increased neural "hyperconnectivity" in the temporal lobe* . . . Maria Konnikova, "From the Words of an Albino, a Brilliant Blend of Color," Literally Psyched (blog), *Scientific American*, February 26, 2013, http://blogs.scientificamerican.com/literally-psyched/2013/02/26/from-the-words-of-an-albino-a-brilliant-blend-of-color.

157 *Most people, "were surprised and disrupted" by these changes to their perception of the world* . . . Stout, *Back to Life*.

157 *A Japanese-American businessman who had an NDE during a motorcycle crash became able to empathetically sense others' emotional pain* . . . Ibid.

157 *Yvonne Kason knew a friend had meningitis before the friend was aware of it* . . . Kason, *Farther Shores*.

158 *Tony Cicoria suddenly gained an obsession with learning to play and compose for the piano after he was struck by lightning* . . . (Panel discussion at the annual meeting of the International Association of Near-Death Studies, Washington, 2013).

158 *In one study, over half* . . . *were found to have "moderate to major" problems taking up the reins of their lives in the aftermath* . . . I. Corbeau, "Psychological Problems and Support After an NDE," *Return: Journal of Near-Death Experiences and Meaning* 15, no. 2 (2004): 16–22.

158 *"It's sad that you can't talk freely about it," one woman said. "I feel penned in. I have so much exuberance and no one to hear me"* . . . Yolaine Stout, *Back to Life* (presentation on aftereffects by psychologist at annual meeting of the International Association of Near-Death Studies, 2008).

159 *Lamented the Spanish mystic Saint John of the Cross: "I no longer live within myself / and I cannot live without God / for having neither him nor myself / what will life be?"* . . . Mandelker, *Pilgrim Souls*, 308, citing "On a Dark Night," *The Collected Works of St. John of the Cross*, trans. Kieran Kavanaugh and Otilio Rodriguez.

159 *One platoon commander described being unable to discipline his troops after his NDE. "All I wanted to do is put my arm around them and say: 'It'll be okay'"* . . . Col. Diane Corcoran (presentation, annual meeting of the International Association of Near-Death Studies, 2008).

160 *An unexpected aspect of combat NDEs is that, because soldiers are often injured together in a blast, they will sometimes go out of body together* . . . Ibid.

160 *There is a variation of this phenomenon called the shared death experience* . . . See, for instance, Glynnis Howarth, "Shared Near-Death and Related Illness Experiences: Steps on an Unscheduled Journey," *Journal of Near-Death Studies* 20, no. 2 (Winter 2001): 71–85.

160 *The psychologist Joan Borysenko described having such an experience when her eighty-one-year-old mother died in the Beth Israel Deaconess Medical Center in Boston* . . . Joan Borysenko (presentation at Omega Center, Rhinebeck, New York, October 2011).

161 *"I felt the joy of his release* . . . Moody, *Glimpses of Eternity.*

161 *"This trait is difficult to describe," Moody writes, "because it takes so many different shapes"* . . . Ibid.

161 *Another frequent perception is the sound of exquisite music* . . . Ibid.

161 *For example, this account from a woman in Virginia: "The night Jim died, I was sitting next to him, holding his hand, when we both left our bodies and began to fly through the air"* . . . Ibid.

162 *In 2013, the late film critic Roger Ebert wrote about his wife's intimation that he was still vital and present during an episode of cardiac arrest* . . . R. Ebert, "I Do Not Fear Death," *Salon*, September 15, 2011.

162 *A cardiac arrest victim was found lying comatose and cyanotic in a field* . . . Lommel, *Consciousness Beyond Life*, 21.

163 *Eighty percent of those who had no memory of an NDE* . . . *made at least one major error in their descriptions, whereas none of the NDE patients made errors* . . . M. Sabom, *Recollections of Death: A Medical Examination* (New York: Simon & Schuster, 1982).

163 *His findings were replicated by doctoral researcher Penny Sartori in 2008 with a study of hospitalized intensive care patients* . . . P. Sartori et al., "A Prospectively Studied Near-Death Experience with Corroborated Out-of-Body Perceptions and Unexplained Healing," *Journal of Near-Death Studies* 27, no. 1 (2008). It is estimated that 10 percent of the general population and perhaps up to 25 percent of children have had a spontaneous sensation of being outside their bodies. This mirrors the percentage split between children who have sensed presences and adults who have, suggesting that consciousness is less firmly entrained in the body in childhood, or that children are more fantasy-prone, depending upon your point of view. Lommel, *Consciousness Beyond Life*, 77.

163 *Sabom found that the NDE group "related accurate details of idiosyncratic*

or unexpected events during their resuscitations" . . . Sabom, *Recollections of Death: A Medical Examination.* See critical discussion in J. M. Holden, "Veridical Perception in Near-Death Experiences," in Holden, *The Handbook of Near-Death Experiences,* 185–212.

163 *Said one: "I was above myself looking down. They was [sic] working on me trying to bring me back. 'Cause I didn't realize at first that it was my body"* . . . Sabom, *Recollections of Death.*

163 *A patient described leaving his body and watching the cardiac surgeon "flapping his arms as if trying to fly"* . . . E. W. Kelly et al., "Can Experiences Near Death Furnish Evidence for Life After Death?" *Omega* 40, no. 4 (2000): 513–19.

164 *In a recent review of ninety-three published reports of potentially verifiable OBE perceptions, 43 percent were found to have been corroborated to the investigator by an independent informant* . . . Holden, "Veridical Perception in Near-Death Experiences," 185–212. Holden reserves judgment on what these reports lead us to conclude, wary of being definitive until there are better-designed and controlled investigations. "The sheer volume of AVP anecdotes that a number of different authors over the course of 150 years have described suggests that AVP is real," she notes. But she believes that "if the controlled investigation of AVP (apparent veridical perception) is to continue, it will likely be a complex and protracted—and therefore, costly—process," p. 197. The Horizon Research Foundation has been funding such efforts in participating cardiac wards in the United States and Europe. Known as the AWARE Study, the wards (ER or cardiology) have placed boards with random symbols high up, facing the ceiling. Parnia discusses the research objectives in his 2013 book, *Erasing Death.*

164 *"The OBE phenomenon is interesting," says the science journalist Jeff Warren, "because you get these incredibly detailed perceptions happening in people with apparently no brain activity"* . . . Jeff Warren, in conversation with the author.

164 *Sam Parnia, director of cardiopulmonary resuscitation research at SUNY in Stony Brook, New York, began investigating NDEs in the context of cardiac resuscitation, and soon heard accounts from his own medical colleagues* . . . Parnia, *Erasing Death,* 228.

166 *Since then, playing God and raising Lazarus has become an increasingly*

widespread...Pim van Lommel, "About the Continuity of Our Consciousness," in *Brain Death and Disorders of Consciousness,* ed. Calixto Machato et alia, New York: Plenum, 2004).

168 *Somehow, in the midst of her induced clinical brain death, Reynolds experienced herself as awake and out of her body . . .* Case summarized by M. Sabom, *Light and Death: One Doctor's Fascinating Account of Near Death Experiences* (Grand Rapids, MI: Zondervan, 1998).

168 *One analysis of the medical records of people reporting NDEs found that they described enhanced mental functioning significantly more often when they were actually physiologically close to death . . .* J. E., Owens et al. "Features of 'Near-Death Experience' in Relation to Whether or Not Patients Were Near Death," *Lancet* 336 (1990): 1175–77.

168 *She also heard the doctors playing "Hotel California," and she joked that the line "you can check out anytime you want but you can never leave" was "incredibly insensitive" . . . The Day I Died: The Mind, the Brain and Near-Death Experiences,* directed by Nick Broome (BBC, 2002).

168 *Reynolds's neurosurgeon, Robert Spetzler, would later say to the BBC: "I don't think the observations she made were based on what she experienced as she went into the operating theater" . . .* Ibid.

169 *Keith Augustine, a young philosopher and avowed skeptic, reviewed the Reynolds case and argued that she could have seen things later and retroactively described them . . .* K. Augustine, "Does Paranormal Perception Occur in Near-Death Experiences?" *Journal of Near-Death Studies* 25, no. 4 (2007): 203–36.

169 *"Ms. Reynolds' brain was being monitored three different ways to ensure that she was deeply and consistently anaesthetized" . . .* Holden, "Veridical Perception in Near-Death Experiences," in Holden, *Handbook of Near-Death Experiences,* 199.

170 *"Memory can supply all the information about your body, what it looks like, how it feels and so on. It can also supply a good picture of the world. 'Where was I? Oh, yes. I was lying in the road after that car hit me'" . . .* Blackmore, *Dying to Live,* 177.

170 *Her critics complain . . .* Carter, *Science and the Near-Death Experience,* 202; also G. Stone, "A Critique of Susan Blackmore's *Dying to Live* and her *Dying Brain Hypothesis.*" http://www.efpf.org.uk/articles/background/dying_to_live. html.

170 *"At last we have a simple theory of the OBE . . . The normal model of reality*

breaks down and the system tries to get back to normal by building a new model from memory and imagination" . . . Blackmore, *Dying to Live.*

170 *Swiss neuroscientist Olaf Blanke . . . suggests that the perception of being out of one's body may be the result of "paroxysmal cerebral dysfunction of the temporo-parietal junction (TPJ)"* . . . O. Blanke et al., "Out-of-Body Experience and Autoscopy of Neurological Origin," *Brain* 127 (2004): 243–58; see also O. Blanke et al., "Stimulating Illusory Own-Body Perceptions: The Part of the Brain That Can Induce Out-of-Body Experiences Has Been Located," *Nature* 419 (2002): 269–70.

171 *"The interpretation of the empirical findings of Blanke and his colleagues is controversial"* . . . Kelly and Kelly, *Irreducible Mind*, 223.

171 *The Austrian neurologist Ernst Rodin, who specialized in epilepsy before his retirement* . . . Rodin, quoted in Kelly and Kelly, *Irreducible Mind*. Position confirmed in email correspondence with author. See also discussion in Holden, *The Handbook of Near-Death Experiences*, 220.

171 *"People are pushing the boundaries to extremes to try to put a label on this," Dr. Sam Parnia wrote* . . . Parnia, *Erasing Death*, 169.

171 *[Parnia] offers the example of a study done with student volunteers, outfitted with special goggles that showed them an image projected by a camera that had been stationed behind them* . . . Ibid., 169. See also Laura Blue, "The Science of Out-of-Body Experiences," *Time*, August 23, 2007, http://www.time.com/time/health/article/0,8599,1655632,00.html.

172 *From this, Parnia complains, "The researchers concluded that they had reproduced an out-of-body experience in the laboratory"* . . . Parnia, *Erasing Death*, 170.

172 *Parnia has found in his own ongoing research in several collaborating hospitals in the United States and Europe that OBEs are being reported by around 2 percent of the cardiac arrest survivors* . . . Sam Parnia, email correspondence with the author. (Parnia has had two OBE cases in his hospital during the AWARE study so far, but neither took place on wards with the boards present. One woman, he says, saw a nurse that she'd never met before, 255.

172 *Cells in the hippocampus, the area of the brain considered crucial for the formation of memory, are particularly sensitive to damage* . . . Vriens et al., "The Impact of Repeated Short Episodes of Circulatory Arrest on Cerebral Function," *EEG Clinical Neurophysiology* 98 (1996): 236–42.

173 *If the dying brain theory were correct, then I would expect that as the oxygen levels in patients' blood dropped, they would gradually develop the illusion of seeing a tunnel and/or a light. In practice, patients with low oxygen levels don't report seeing a light, a tunnel, or any of the typical features of an NDE"* . . . Parnia, What Happens When We Die, 21.

173 *A near-death experience is something more akin to a waking dream* . . . Nelson, The Spiritual Doorway in the Brain.

173 *Nelson has publically described NDEs as "spiritual experiences," but he thinks they can be scientifically accounted for* . . . See, for example, Kevin Nelson, interview with Alex Tsarkiris, Skeptiko, January 27, 2010, http://www.skeptiko.com/kevin-nelson-skeptical-of-near-death-experience-accounts.

174 *The radiologist Jeffrey Long and the psychologist Janice Miner Holden responded to Nelson that 40 percent of the NDEers in their survey had actually said "no" to questions* . . . J. Long and J. Holden, "Does the Arousal System Contribute to Near-Death and Out-of-Body Experiences?: A Summary and Response," *Journal of Near-Death Studies* 25, no. 3 (2007).

174 *The Brown University researcher Willoughby Britton found—in an unrelated study—that, after their NDE, people entered REM sleep almost sixty minutes later, on average, than other sleepers* . . . Britton and Bootzin, "Near-Death Experiences and the Temporal Lobe."

174 *"I make my living as a clinical psychiatrist dealing with people who have trouble dealing with what's real and what's not real, what's not delusional," says the psychiatrist Bruce Greyson* . . . Bruce Greyson (United Nations symposium on the launch of the Human Consciousness Project, New York, September 11, 2008).

174 *When Nancy Evans Bush, who had a frightening NDE, was asked what question she found most irritating to be asked, she said, "Probably the one I dislike the most is 'Do you believe these NDEs?'"* . . . Nancy Evans Bush, interview by Amy Stringer, "Reflections from Three Decades with IANDS," *Vital Signs* 28, no. 4 (Fall 2009).

175 *As Bruce Greyson said in 2008 at the UN-sponsored meeting on the mind-body problem in New York, where he mused about the nature of reality, "We have no blood levels for enlightenment, but we can study its aftereffects"* . . . (U.N. symposium, New York, September 11, 2008).

175 *The percentage of Australians in Cherie Sutherland's study who claimed to have no religious affiliation prior to their NDE was 46 percent; that number jumped*

to 84 percent afterward . . . C. Sutherland, *Transformed by the Light: Life After Near-Death Experiences* (Sydney: Bantam, 1992).

175 *Pim van Lommel's Dutch research showed that church attendance declined by 42 percent in his NDE group, while spirituality nearly doubled* . . . Lommel, *Consciousness Beyond Life*, 57.

176 *Compassion and empathy for others increased by 73 percent at the eight-year mark versus 50 percent for the control group in Lommel's study. Interest in material status decreased by 50 percent; whereas that same interest—the pursuit of money, the quest for accomplishment—increased by 33 percent in controls* . . . Ibid.

176 *It takes someone who has had a deep NDE an average of twelve years to integrate the "radical shift in reality" into their lives* . . . Stout, *Back to Life*.

176 *[David Bennett] "It was like being hit by a two-by-four," he told me, "and I was like, okay, okay, I get it. Now I'm going to have to start living my life knowing what I know"* . . . David Bennett, interview with the author.

177 *[Eben Alexander] "Time flow in that realm is very different, so I don't know how long I was in a murky underground place"* . . . Eben Alexander (lecture Syracuse, New York, May, 2013).

179 *Confidently, Mum quoted an ancient Roman thinker: "Where death is, I am not; where I am, death is not"* . . . Epicurus. See, for example, Simon Critchley, *Book of Dead Philosophers* (New York: Vintage, 2009).

179 *I am fonder of this quotation, commonly attributed to the Indian Nobel Laureate for Lieterature in 1913, Rabin-dranath Tagore: "Death is not extinguishing the light; it is only putting out the lamp because the dawn has come"* . . . This is widely attributed to Tagore, but I haven't been able to find it in his work. Instead, I have found it in the much humbler source of an American Catholic priest named Floyd Lotito, who wrote a short book of inspirational thoughts. Floyd Lotito, *Wisdom, Age and Grace: An Inspirational Guide to Staying Young at Heart* (New Jersey: Paulist Press, 1993).

WALKING THE ENCHANTED BOUNDARY

181 *"The more that critical reason dominates, the more impoverished life becomes"* . . . Jung, *Memories, Dreams, Reflections*, 302.

183 *[Raymond Moody] "Her face was going gray," he recalls of one patient he was trying to resuscitate in medical school, "my resuscitation didn't work, and yet she was reciting poetry"* . . . Raymond Moody (lecture, Omega Center, Rhinebeck, NY, October 2011).

190 *"Where is the wisdom we have lost in knowledge? Where is the knowledge we have lost in information?"* . . . Eliot, "Choruses from 'the Rock.'"

192 *Nancy Evans Bush, who spent years contemplating her frightening near-death experience, writes, "We have to recognize that the universe is made up of darkness as well as light"* . . . N. Bush, "Distressing Western Near-Death Experiences: Finding a Way through the Abyss," in Holden, *The Handbook of Near-Death Experiences,* 87–108.

192 *"The deepest enigma for human beings [remains] learning to live with what we believe. That's the hard part"* . . . Ibid.

194 *When the philosopher Ken Wilber lost his new wife to cancer, she kept telling him repeatedly as she was dying, "Promise me that you'll find me"* . . . Wilber, *Grace and Grit.*

194 *"This is in the end the only kind of courage that is required of us: the courage to face the strangest, most unusual, most inexplicable experiences that can meet us"* . . . Rainer Maria Rilke, *Letters to a Young Poet.*

Bibliography

Arcangel, Dianne. *Afterlife Encounters: Ordinary People, Extraordinary Experiences*. Charlottesville, Va: Hampton Roads Publishing, 2005.

Aries, Philippe. *The Hour of Our Death*. New York: Vintage Books, 2008.

Bachelard, Gaston. *The Poetics of Reverie: Childhood, Language and the Cosmos*. Boston: Beacon Press, 1971.

Barrett, Sir William. *Deathbed Visions: How the Dead Talk to the Dying*. Guildford, UK: White Crow Books, 2011.

Beauregard, Mario, and Denyse O'Leary. *The Spiritual Brain: A Neuroscientist's Case for the Existence of the Soul*. San Francisco: HarperOne, 2007.

Bennett, David. *Voyage of Purpose: Spiritual Wisdom from Near-Death Back to Life*. Forres, Scotland: Findhorn Press, 2011.

Blackmore, Susan. *Dying to Live: Science and the Near-Death Experience*. London: Grafton Books, 1993.

Bolte-Taylor, Jill. *My Stroke of Insight: A Brain Scientist's Personal Journey*. New York: Plume, 2009.

Callanan, Maggie, and Patricia Kelley. *Final Gifts: Understanding the Special Awareness, Needs, and Communications of the Dying*. New York: Bantam, 1992.

Cardeña, Etzel, Steven Jay Lynn, and Stanley Krippner, eds. *Varieties of Anomalous Experience: Examining the Scientific Evidence*. Washington: American Psychological Association, 2000.

Carter, Chris. *Science and the Afterlife Experience*. Rochester, VT: Inner Traditions, 2012.

———. *Science and the Near-Death Experience*. Rochester, VT: Inner Traditions, 2010.

Coffey, Maria. *Explorers of the Infinite: The Secret Spiritual Lives of Extreme Athletes—and What They Reveal About Near-Death Experiences, Psychic Communication, and Touching the Beyond*. New York: Tarcher, 2008.

Davis, Wade. *The Wayfinders: Why Ancient Wisdom Matters in the Modern World.* Toronto, ON: House of Anansi Press, 2009.

Ecklund, Elaine Howard. *Science vs. Religion: What Scientists Really Think.* London: Oxford University Press, 2010.

Eliot, T. S. "Choruses from 'the Rock.'" *Collected Poems, 1909–1935.* New York: Harcourt, Brace and Company, 1934.

———. *Prufrock and Other Observations.* London: The Egoist, 1917. Reprinted by Kessinger Publishing, 2010.

———. *The Waste Land.* Edited by Michael North. New York: W. W. Norton & Company, 2000.

Fenwick, Peter, and Elizabeth Fenwick. *The Art of Dying.* London: Continuum Books, 2008.

Figes, Orlando. *Just Send Me Word: A True Story of Love and Survival in the Gulag.* New York: Metropolitan Books, 2012.

Furlong, Monica. *Visions and Longings: Medieval Women Mystics.* Boston: Shambhala, 1996.

Geiger, John. *The Third Man Factor: The Secret to Survival in Extreme Environments.* Toronto, ON: Penguin Canada, 2009.

Gratton, Nicole, and Monique Séguin. *Les rêves en fin de vie: 100 récits de rêves pour faciliter la grande traversée.* Québec: Flammarion, 2009.

Grof, Stanislav. *Spiritual Emergency: When Personal Transformation Becomes a Crisis.* New York: Tarcher, 1989.

Gurney, Edmund, Frederic W. H. Myers, and Frank Podmore. *Phantasms of the Living.* 2 vols. London: Truber, 1886.

Hamilton, Allan. *The Scalpel and the Soul: Encounters with Surgery, the Supernatural and the Healing Power of Hope.* New York: Tarcher/Penguin, 2008.

Haraldsson, Erlendur. *The Departed Among the Living: An Investigative Study of Afterlife Encounters.* Guildford, UK: White Crow Books, 2012.

Hawker, Paul. *Secret Affairs of the Soul: Ordinary People's Extraordinary Experiences of the Sacred.* Kelowna: Northstone Publishing, 2000.

———. *Soul Survivor: A Spiritual Quest through 40 Days in the Wilderness.* Sydney: Lion Books, 2001.

Holden, Janice Miner, Bruce Greyson, and Debbie James, eds. *The Handbook of Near-Death Experiences: Thirty Years of Investigation.* Santa Barbara, CA: Praeger Publishers, 2009.

Horgan, John. *Rational Mysticism: Spirituality Meets Science in the Search for Enlightenment.* New York: Houghton Mifflin, 2003.

Bibliography

Hufford, David. *The Terror That Comes in the Night: An Experience-Centered Study of Supernatural Assault Traditions*. Philadelphia: University of Pennsylvania Press, 1989.

Huxley, Aldous. *The Doors of Perception*. New York: Harper, 1954.

James, William. *The Varieties of Religious Experience*. New York: The Modern Library, 1902.

Jorgensen, Rene. *The Light Behind God: What Religion Can Learn From Near Death Experiences*. Montréal: CreateSpace Independent Publishing Platform, 2011.

Jung, C. G. *Memories, Dreams, Reflections*. Recorded and edited by Aniela Jaffé. Translated from the German by Richard and Clara Winston. New York: Vintage, 1989.

———. *Psychology and Religion*. New Haven: Yale University Press, 1938.

Kason, Yvonne. *Farther Shores: Exploring How Near-Death, Kundalini and Mystical Experiences Can Transform Ordinary Lives*. Toronto, ON: HarperCollins, 1994.

Kellehear, Allan. *Experiences Near Death: Beyond Medicine and Religion*. London: Oxford University Press, 1996.

Kelly, Edward, and Emily Williams Kelly, eds. *Irreducible Mind: Toward a Psychology for the 21st Century*. Plymouth, UK: Rowman & Littlefield, 2007.

Kelly, Emily Williams. *Science, the Self and Survival after Death: Selected Writings of Ian Stevenson*. Plymouth, UK: Rowman & Littlefield, 2012.

Kessler, David. *Visions, Trips and Crowded Rooms*. Carlsbad, CA: Hay House, 2010.

Kircher, Pamela. *Love Is the Link: A Hospice Doctor Shares Her Experience of Near-Death and Dying*. Pagosa Springs, CO: Awakening Press, 2013.

Kübler-Ross, Elisabeth. *On Life after Death*. Toronto, ON: Celestial Arts, 2008.

Kuhl, David. *What Dying People Want: Practical Wisdom for the End of Life*. Toronto, ON: Anchor Canada, 2003.

Kuhn, Thomas. *The Structure of Scientific Revolutions*. 50th Anniversary Edition. Chicago: University of Chicago Press, 2012.

Lagrand, Louis. *Love Lives On*. New York: Berkley Books, 2006.

Lecouteux, Claude. *The Return of the Dead: Ghosts, Ancestors, and the Transparent Veil of the Pagan Mind*. Rochester, VT: Inner Traditions, 2009.

Lommel, Pim van. *Consciousness Beyond Life: The Science of the Near-Death Experience*. San Francisco: HarperOne, 2010.

Mandelker, Amy, and Elizabeth Powers, eds. *Pilgrim Souls: A Collection of Spiritual Autobiography*. New York: Touchstone, 2000.

Mayer, Elizabeth Lloyd. *Extraordinary Knowing: Science, Skepticism and the Inexplicable Powers of the Human Mind.* New York: Bantam, 2007.

McLuhan, Robert. *Randi's Prize: What Sceptics Say about the Paranormal, Why They Are Wrong and Why It Matters.* Leicester, UK: Matador, 2010.

Meyers, F. W. H. *Human Personality and Its Survival of Bodily Death.* London: Longmans Green, 1903.

Moody, Raymond. *Glimpses of Eternity: Sharing a Loved One's Passage from This Life to the Next.* New York: Guideposts, 2010.

———. *Life After Life: The Investigation of a Phenomenon—Survival of Bodily Death.* Originally published in 1975. San Francisco: HarperOne, 2001.

Nelson, Kevin. *The Spiritual Doorway in the Brain: A Neurologist's Search for the God Experience.* New York: Dutton, 2012.

Newberg, Andrew, and Eugene d'Aquili. *Why God Won't Go Away: Brain Science and the Biology of Belief.* New York: Ballantine Books, 2001.

Oliver, Mary. *A Thousand Mornings.* Reprint edition. New York: Penguin Books, 2013.

Osis, Karlis, and Erlendur Haraldsson. *At the Hour of Death: A New Look at Evidence for Life After Death.* Revised edition. Mamaroneck, NY: Hastings House, 1990.

Otto, Rudolf. *The Idea of the Holy: An Inquiry into the Non-rational Factor in the Idea of the Divine and its Relation to the Rational.* Middlesex: Penguin Books, 1959.

Parnia, Sam. *Erasing Death: The Science That Is Rewriting the Boundaries between Life and Death.* HarperOne, San Francisco: 2013

———. *What Happens When We Die: A Groundbreaking Study into the Nature of Life and Death.* Carlsbad, CA: Hay House, 2006.

Perera, Mahendra, Karuppiah Jagadheesan, and Anthony Peake, eds. *Making Sense of Near-Death Experiences: Handbook for Clinicians.* London: Jessica Kingsley Publishers, 2011.

Radin, Dean. *Entangled Minds: Extrasensory Experiences in a Quantum Reality.* New York: Pocket Books, 2006.

———. *Supernormal: Science, Yoga, and the Evidence for Extraordinary Psychic Abilities.* New York: Deepak Chopra Books, 2013.

Rilke, Rainer Maria. *Duino Elegies.* New York: Vintage, 2009.

Rinpoche, Sogyal. *The Tibetan Book of Living and Dying.* San Francisco: Harper, 1993.

Bibliography

Roach, Mary. *Spook: Science Tackles the Afterlife.* New York: Norton, 2005.

Roper, Michael. *The Secret Battle: Emotional Survival in the Great War.* Manchester, UK: Manchester University Press, 2009.

Sacks, Oliver. *Hallucinations.* New York: Vintage, 2013.

———. *Musicophilia: Tales of Music and the Brain.* Toronto, ON: Vintage Canada, 2007.

Sagan, Carl. *Contact.* New York: Doubleday, 1997.

Seaberg, Marueen. *Tasting the Universe: People Who See Colors in Words and Rainbows in Symphonies; A Spiritual and Scientific Exploration of Synesthesia.* Pompton Plains, NJ: New Page Press, 2011.

Sheldrake, Rupert. *Dogs That Know When Their Owners Are Coming Home.* Revised edition. New York: Broadway Books, 2011.

———. *The Science Delusion: Freeing the Spirit of Scientific Inquiry.* London: Coronet, 2012.

Singh, Kathleen Dowling. *The Grace in Dying: a Message of Hope, Comfort and Spiritual Transformation.* San Francisco: HarperOne, 2000.

Slocum, Joshua. *Sailing Alone Around the World.* Edited by George Stade. New York: Barnes and Noble Classics, 2005.

Stevenson, Ian. *Telepathic Impressions: A Review and Report of 35 New Cases.* Charlottesville, VA: University of Virginia Press, 1973.

Tart, Charles. *The End of Materialism: How Evidence of the Paranormal Is Bringing Science and Spirit Together.* Oakland, CA: New Harbinger Publications, 2012.

Taylor, Rob. *The Breach: Kilimanjaro and the Conquest of Self.* Wildeyes Press, 1991.

Taylor, Timothy. *The Buried Soul: How Humans Invented Death.* Boston: Beacon Press, 2002.

Terkel, Studs. *Will the Circle Be Unbroken: Reflections on Death, Rebirth and Hunger for a Faith.* New York: Ballantine Books, 2001.

Tompkins, Ptolemy. *The Modern Book of the Dead: A Revolutionary Perspective on Death, the Soul, and What Really Happens in the Life to Come.* New York: Atria Books, 2012.

Underhill, Evelyn. *Mysticism: The Preeminent Study in the Nature and Development of Spiritual Consciousness.* Originally published in 1911. London: Kessinger Publishing, 2010.

Warren, Jeff. *The Head Trip: Adventures on the Wheel of Consciousness.* Toronto, ON: Random House, 2007.

Bibliography

Wilber, Ken. *Grace and Grit: Spirituality and Healing in the Life and Death of Treya Killam*. Boston: Wilber Shambhala Press, 2000.

Zaleski, Carol. *Other World Journeys: Accounts of Near-Death Experience in Medieval and Modern Times*. London: Oxford University Press, 1987.

Zimmer, Carl. *Soul Made Flesh: The Discovery of the Brain and How It Changed the World*. New York: Atria, 2005.

Index

About the Author

Patricia Pearson is a journalist, novelist, and humorist whose work has appeared in the last year in the *New Yorker*, *New York Times*, *Huffington Post*, and *BusinessWeek*, among other publications. A longtime member of *USA Today*'s op-ed Board of Contributors, she directed the research for the 2009 History Channel documentary *The Science of the Soul*. She has appeared on NPR and other media, and her recent TED talk was titled "Why Ghosts Are Good for You," pointing to the research that shows how important experiences with unexplained phenomena can be in helping people cope with their grief. She is based in Toronto.